Design for Performative Arts Spaces

Alessandro Melis · Marco Pelle
Editors

Design for Performative Arts Spaces

Historical Evolution, Cultural Context, and Future Opportunities

Springer

Editors
Alessandro Melis
School of Architecture and Design
New York Institute of Technology
New York, NY, USA

Marco Pelle
Department of Social and Political Sciences
Bocconi University
Milan, Italy

ISBN 978-3-031-98214-9 ISBN 978-3-031-98215-6 (eBook)
https://doi.org/10.1007/978-3-031-98215-6

This Springer imprint is published by the registered company Springer Nature Switzerland AG
The registered company address is: Gewerbestrasse 11, 6330 Cham, Switzerland

Acknowledgments

We extend our deepest appreciation to the IDC Foundation for its steadfast support, which has been crucial to the success of the practice-based research projects showcased in this book. We are particularly grateful to Raymond Savino and Henry Miller for their exceptional support.

Our sincere thanks also goes to Mariangela Zappia, Ambassador of Italy to the USA.

We are indebted to Renzo Macelloni, Mayor of Peccioli, and Arianna Merlini, General Director of Belvedere SpA; Maria Porro, President of Salone del Mobile Milano; Marva Griffin, the visionary behind Salone Satellite; and Patrizia Catalano and Maurizio Barberis of Hoperaperta, for their invaluable partnership in the exhibitions that highlighted practice-based research at the School of Architecture and Design at the New York Institute of Technology.

We would also like to express our heartfelt gratitude to the following individuals, whose insights, collaboration, and support have been indispensable to our work: Roberto Frangione, Eva Sohlman, Neal MacFarquhar, Nancy Pena, Telmo Pievani, Heather Pringle, Ian Tattersall, Niles Eldredge, Elisabeth Vrba, Richard Sennett, Wolf D. Prix, Thomas Auer, Walter Mayrhofer, Luca Boy, Simone Sfriso, Jesus D'Alessandro Martinez, David Turnbull, Walter Mayrhofer, Consuelo Nava, Mose' Ricci, Paolo Belardi, Monica Battistoni, Domenico Lucanto, Christian Pongratz, Pamela Cole, Giovanni Pellerito, Elisabeth Aitken Rose, Julia Gatley, Chris Chang, Trevor Keeble, Tarek Teba, Amy Bravo, Liam Stumbles, Jumhur Gokchepenar, Emanuele Lisci, Gian Luigi Melis, Nico Panizzi, Ilaria Fruzzetti, Filippo Mariani, Donatella Ulzega, Gian Benedetto Melis, Luca Gregorio, Francesco Secchiaroli, Eric Goldemberg, Daniela Perrotti, Paola Boarin, Malina Mannarino, and William Carpenter.

A special note of appreciation goes to Natalia Giovannetti Alves Feitosa for her integral role in the editorial organization of this book.

Their contributions—whether through inspiration, expertise, collaboration, guidance, or support—have been instrumental in the realization of this work. We are deeply thankful for their dedication and involvement.

Contents

Part III Case Studies and Best Practices

Part I
Foundational Frameworks of Performative Art

The initial part of this book lays a robust foundation by defining essential terms related to performative art, elucidating their meanings and origins, and situating them within their historical, cultural, philosophical, and social contexts. It also traces the evolution of physical spaces dedicated to performative arts.

Chapter 1, "Understanding Performative in the Context of Arts: Definitions, Categories, Taxonomies, and Classifications," examines the evolution of performative arts through a multidisciplinary lens. This exploration broadens traditional taxonomies and investigates how artistic expressions, ranging from cultural rituals to political activism, function as performative acts. By challenging conventional classifications, the chapter proposes a more nuanced understanding of art's role in human creativity and societal transformation.

Chapter 2, "From Creativity to Evolutivity," creativity is explored as a force transcending individual and temporal boundaries, acting as a mechanism of cultural and societal evolution. By incorporating principles of evolutionary biology, such as Gould and Eldredge's Punctuated Equilibrium, the chapter reveals how creativity mirrors rapid, transformative evolutionary leaps. Embracing new discoveries, this chapter delves into art's role in human development, proposing the concept of "evolutivity" to describe creativity's evolutionary impact.

Chapter 3, "Origins of Performance," explores how performative arts like theater, dance, music, and spoken word exemplify "evolutivity" by integrating live performances and audience interaction, embodying life's transient nature. It addresses the blurred lines between recorded and live performances, especially in cinema, and examines the historical roots and evolutionary timeline of performative arts, considering cognitive development and abstract thinking. The chapter also investigates prehistoric art, burial rituals, and petroglyphs as early forms of performative expression, linking them with evolutionary biology and epistemology to highlight their role in shaping human experience.

The final chapter of the first part, "Architectural Space for Performative Arts," explores how performative arts are deeply rooted in Aristotle's theory of mimesis, as established in "Poetics" and "Metaphysics." This chapter traces the evolution of performance spaces, from ancient Greek and Roman theaters to medieval, Renaissance, and modern venues, emphasizing their role in shaping and reflecting societal values and artistic expressions. The chapter also explores avant-garde movements and contemporary theater architecture. Venues like the Centre Pompidou and Walt Disney Concert Hall exemplify innovation, flexibility, and technological advancements, ensuring the dynamic evolution of performative arts.

Chapter 1
Performative in the Context of Arts: Definitions, Categories, Taxonomies, and Classifications

Alessandro Melis and Marco Pelle

Abstract This chapter explores "performative" in the arts, tracing its roots from linguistic theory to expansive artistic applications. From theater and dance to live performances, performative arts emphasize interactivity, often blurring lines between traditional categories. By investigating evolving taxonomies informed by fields like evolutionary biology and paleoanthropology, the chapter argues for fluid classifications that embrace complex realities, challenging static perceptions in art history, and introduces concepts like exaptation to reveal how performative acts—spontaneous or scripted—emerge as social tools and collective expressions beyond mere functionality. Highlighting art's associative, non-linear nature, the chapter contemplates how performative acts shape cultures and societies. Ultimately, it posits performative arts as vital for expanding collective creativity and reimagining the intersections of environment, society, and individual agency in contemporary art practice.

Keywords Performative arts · Linguistic theory · Interactivity · Artistic applications · Theater and dance · Live performances · Evolving taxonomies · Evolutionary biology · Paleoanthropology · Fluid classifications · Art history · Exaptation · Social tools · Collective expression · Non-linear nature · Cultural shaping · Collective creativity · Environmental intersections · Individual agency · Contemporary art practice

Performativity denotes the idea that language acts as a form of social action capable of initiating change. John L. Austin, a philosopher of language, introduced this concept in his influential work "How to Do Things with Words" (1962), emphasizing the unique ability of speech to perform actions. He distinguished performative

A. Melis (✉)
School of Architecture and Design, New York Institute of Technology, New York, USA
e-mail: amelis@nyit.edu

M. Pelle
Department of Social and Political Sciences, Bocconi University, Milano, Italy

A. Melis and M. Pelle (eds.), *Design for Performative Arts Spaces*,
https://doi.org/10.1007/978-3-031-98215-6_1

language from constative language, which merely describes facts and can be evaluated as true or false. Examples of performative language include making promises, placing bets, conducting wedding ceremonies, an umpire declaring a foul, or a judge delivering a verdict. The concept of "performative" extends beyond mere performance. It embodies the idea that the act itself accomplishes something in the world. Austin (1962) introduced "performative utterances" to describe statements that do not merely convey information but perform an action.

From its origins in linguistic theory, the term "performative" has broadened to various disciplines, particularly the arts, where it captures multiple dimensions of artistic expression beyond static notions. Performative arts emphasize action, process, and engagement, highlighting the interactive, experiential, and often ephemeral nature of artistic endeavors. In performative arts, as in linguistic theory, the act of performance is not merely an artistic expression but a transformative experience for both the performer and the audience (Butler, 1990).

Performance art is a primary category where performativity is central, involving live presentations where the artist's actions, presence, and interactions with the audience are crucial. This genre includes a wide range of art forms executed in the presence of an audience, utilizing the body, voice, or objects to convey artistic expression. Unlike static art forms like painting or sculpture, performative arts are ephemeral, dynamic, and interactive, creating a unique, transient experience for both the performer and the audience.

Theater, dance, music, and performance art all fall under the umbrella of performative arts. In theater and drama, the performative element is inherent, relying on live enactment, dialogue, and physical expression to convey narratives and emotions. Dance uses the human body as a medium of expression, with movements, choreography, and spatial interactions contributing to its performative nature.

These art forms are characterized by their live, interactive nature, where the audience's presence plays a critical role in the experience and often in the creation of the art itself (Goldberg, 1979). The immediacy and temporality of performative arts distinguish them from other art forms, imbuing them with a potent capacity for emotional impact and social commentary.

Although starting from these aforementioned conventional categories, this book doesn't intend to provide a comprehensive understanding of its role in contemporary artistic practice. Instead, it aims to explore the possibility of extending current taxonomies through a multidisciplinary approach, towards the blurring confines where human activities and art performance start merging.

The book also argues that cultural and religious rituals, dating back to ancient times, can be considered forms of embryonic performances that integrate traditional practices and symbols, often blurring the line between art, ceremony, and life. Consistently, activism and social commentary can be seen as performative acts with a political dimension, aiming to address social issues, provoke thought, and inspire change. Site-specific performances, designed for particular locations and integrating the environment into the artistic experience, are closely tied to the spatial context.

Similarly, interactive and participatory art includes works that actively involve the audience, inviting them to participate and influence the outcome of the art. The performative aspect is realized through the engagement and responses of the participants. These examples show that performative arts can also be classified based on their mode of presentation. Live performances involve direct interaction with the audience, while mediated performances are transmitted through digital or recorded media.

The number of participants can also be used as a taxonomy: solo performances focus on a single artist's actions, whereas collaborative performances involve multiple artists interacting and creating together. Performative arts can also be categorized based on the degree of pre-planning. Scripted performances follow a predetermined script or choreography, while improvised performances are created spontaneously. Conceptual performances emphasize ideas and concepts over traditional aesthetics, with the performative act serving as a vehicle to explore and communicate abstract notions.

Without neglecting traditional categories, classifications, and taxonomies, the book aims to extend and potentially recombine them through the lens of recent discoveries in fields such as paleoanthropology, archeology, and evolutionary biology.

When discussing creativity, traditional literature primarily focuses on art, its definition—what art ontologically is—and its categories (performative arts, fine arts, visual arts, architecture, etc.). These categories further develop into classifications (for instance, within performing arts: theater, music, dance) and extend to chronological and geographical taxonomies within the hierarchical scope of what we define as history. History, in its most conventional sense, is the study of human-related past events, including the rise and fall of civilizations, political developments, cultural transformations, economic changes, and social movements. This latter specification is so obvious as to sound banal and redundant, and this is exactly the point the authors of this book want to make: taxonomies tend to crystallize over time, being perceived as self-evident, while they also risk providing a limited perspective of reality here described as the realm of possibilities.

Reifications, for instance, are abstractions of reality that might seem self-evident because we assume them as facts. This is the case, for instance, with human history as a separate discipline from paleoanthropology, marked by the alleged cesura of the birth of civilization around 10,000 years ago. This is a taxonomy, not a fact. We argue here that other taxonomies are possible, linking art to human creativity in a way different from a linear progression along the chronological axis of history. Alternative perspectives might not result in alternative taxonomies; instead, they might favor an extension of taxonomy, open to a more complex reality, which could become a valuable reservoir of possibilities in times of global crises (Melis et al., 2024).

Categories and classifications were originally intended to organize items or concepts into nested levels of increasing specificity, providing a comprehensive framework for understanding the relationships and hierarchies among different items. However, considering more recent discussions about the extension of traditional taxonomies, we realize that orthodox classification might fail to describe higher levels of complexity, with the risk of marginalizing items on the periphery of the taxonomical tree, for instance, as in the case of exaptation in evolutionary biology.

Particularly regarding creativity, which by definition is fed by associative thinking, overlooking the relationships among classified items located in different and distant branches of that tree might pose a higher risk of undermining the possibility of taxonomical nuances and adjacencies. These nuances and adjacencies can enrich the spectrum of possibilities to describe and understand how art, creativity, and human cognitive capabilities are related and their potential.

While the purpose of these categories, classifications, and taxonomies is to create a structured framework that aids in the identification, naming, and grouping of items, making it easier to study, compare, and analyze them, they also risk missing associative thinking opportunities that might maximize human potential for artistic expression or a deeper understanding of its manifestations.

Simply put, creativity can be defined as the cognitive capability to build relationships beyond organized classification. This is why taxonomies, when intended as a reductionist form of hierarchical order, will not suffice to describe what art can be. In moments when artistic expressions do not allow us to distinguish the musical components from the movement components of dance, their architectural significance in terms of space, or what all these interactions might signify from a scientific perspective or more holistically, taxonomies fall short.

When did we start distinguishing intuitions, such as those suggesting the Theory of Relativity, as separate from art and performance? On the more practical end of the spectrum, is recognizing the utilitarian impact of an engine, in transport, enough to classify it as engineering and not an installation? Similarly, how did a "still life" abstract painting transition from being a biological study to an artistic gesture?

Architecture is central to this discussion as criticism often oscillates between valuing its pragmatic use and its fine arts aspect, yet rarely considers its performative potential. Archeological evidence may suggest, instead, that the ancestral origins of architecture, from ritual burials to star observations, are closely linked to what we would define today as performance.

In all these cases, we can recognize the codified and almost obligatory associations between intentionality, rationality, order and function, beauty, and historical epochs. Through the decoupling these association new reflections and more articulated perspectives on extended and creative taxonomies could emerge.

According to Pievani (2023), the pressure against the order of perfection enables the unexpected byproducts of human thought and action, which we might call art. The persistence of art activities through millennia might therefore be explained by its positive impact, helping humans overcome emergencies and contingencies when rational thinking as a standard mode of survival falls short.

Thus, art involves the capacity to create relationships among diverse ideas that may emerge unintentionally and persist due to feedback phenomena associated with aesthetic pleasure, which, according to the concept of exaptation, also shows that non-intentionality has a functional role in evolutionary terms.

Human reluctance to accept a more complex level of functionality is therefore due to the difficulty in discerning the evident cause and consequence from the inscrutable deep-time evolutionary correlations between the environment, human survival, and cognitive and subliminal feedback phenomena. These complex correlations, in fact, involve more variables than our brains can process rationally and span timescales that often exceed the civilization timespan as defined by historical studies. Accordingly, more simplified taxonomies, possibly binary—for instance, where notions like functionality, rationality, and intentionality overlap—provide a comfort zone.

Evolutionary Perspectives on Performative Arts

Since the Renaissance and until the late eighteenth century, art history focused primarily on the end of the Bronze Age and the origin of the classical world as the main period of cultural transformation. With the inception of archaeological studies and the growing Enlightenment interest in exotism, geographically diverse cultures—from China to Mesopotamia and Egypt—became a new source of inspiration, manifested through revivalist and eclectic art expressions (Melis & Melis, 1996). However, classical centrality persisted in all forms of art, including performative ones, becoming the flagship of the new Western democracies until and beyond the avant-garde movements of the twentieth century.

Criticism of Western-centric classical and modernist perspectives in architecture and art began tentatively in the 1960s (Rudofsky, 1987), although the term "Neolithic Revolution," coined by V. Gordon Childe in his book *Man Makes Himself*, had already indicated a diverse spread of civilization about 12,000 years ago. This period was marked by the birth of agriculture and the Middle East as a crucial region for human settlements and the development of more complex societies (Childe, 1936).

However, recent studies informed by paleoanthropology have supported an even more radical view of civilization, moving the timeline back before the Neolithic Revolution and suggesting that diverse types of complex societies might have existed before the inception of agriculture.

For instance, the mature artistic expressions at Göbekli Tepe, inhabited from around 9500, during the Pre-Pottery Neolithic, have disrupted several associations taken for granted by current taxonomies of art history, favoring renewed cultural and anthropological positions (Graeber & Wengrow, 2021).

If humans were able to build something as complex and symbolically refined as Göbekli Tepe without the framework of a settled organized society as we conventionally understand it, which seemed a conditio sine qua non until recently, how many interpretive mistakes have resulted from this initial misinterpretation? Does the distinction between performative art, fine art, and architecture as expressed in the massive stones of Göbekli Tepe apply in any way to a society that seems to contradict our basic instruments of cultural and anthropological interpretation? And if not, how much should these distinctions rule the way we understand art today, as they might instead, at least in some cases, blind us to a more complex understanding of the arts?

The existence of one or more societies capable of organizing thousands of people around a creative endeavor without being backed by any primitive form of polis—understood as a structure in which an organized built settlement and its political system, defining the rules of negotiation between a society and its government, coincide—challenges both the conventional historical idea of progress and the classification of art and technology (the latter traditionally viewed as having a more pragmatic and rationalistic functional scope).

A similar broadening of perspective occurred in evolutionary biology during the 1980s and in neuroscience in the early 2000s. These advancements helped push back the emergence of human arts to about 40,000 years ago and identified the birth of creative cognitive potential in human brains as occurring around 200,000–150,000 years ago.

Several research gaps arise from these considerations, especially concerning the collective importance of creativity. This aspect has been overshadowed by art history's focus on a chronological sequence of periods, which tends to emphasize the progressive refinement of art and portrays individual artists as epic heroes.

The increasingly culturally and technologically complex works of art superficially seem to corroborate the idea of linear progress, suggesting that the brain of the Renaissance man was uniquely capable of creating art that would have been cognitively impossible for a human living 20,000 years ago. This notion has also resulted, since the nineteenth century, from recurring misinterpretations and the unconsciously cherry-picked aspects of Darwinian theories. For instance, as we uncritically accept the idea that brain size indicates a certain level of increased intelligence and artistic talent, we overlook facts that contradict the narrative of progress, such as the fact that our brains are 13% smaller than those of humans from 100,000 years ago.

The cult of the individual artist has not been fueled by empirical evidence. When the notion of linear progress fails, art historians rely on other recurring paradigms, often supported by consolatory narratives that may even contradict the idea of progress. For instance, art literature is replete with descriptions of Classicism and the Renaissance, with their ideals of absolute beauty and proportion, where one can perceive echoes of a subtle and sophisticated idea of a Golden Age. These periods, as well as the Middle Ages and even modernism in different cultural contexts, are seen as benchmarks for all previous and subsequent art expressions, offering a sense of confidence in humanity's elevated place in history.

If we accept the scientific evidence that the cognitive potential of contemporary humans is not significantly different from that of their early ancestors, we might recognize that the ability of several receptive artists to understand and represent central perspective in the early fourteenth century is more attributable to shared cultural factors than to their individual capabilities. This perspective could stimulate a discussion of art and creativity as a collective potential, which might be studied and fully understood through an extension of art history paradigms.

Let's Dance: Beyond the Dichotomy of Creationism and Orthodox Evolutionism

In reimagining the frameworks that have historically categorized our understanding of the world—such as the creationist notion of a Golden Age or the linear progress envisaged by orthodox evolutionism—we might find value in considering an alternative perspective. This perspective would still grapple with the grand questions of our origin and journey but would embrace the inherent complexity of these inquiries, recognizing that no model can be definitive or perfect.

Thomas Huxley's famous metaphor of evolution as a chess game between humanity and nature illustrates a traditional view where human progress is seen as a series of strategic moves against an opposing force (Gould, 1996). This metaphor, while powerful, is deeply rooted in an orthodox evolutionary approach that often positions humanity in a central or adversarial role to nature, emphasizing a linear progression of improvement.

To align more closely with contemporary scientific discoveries, we could shift our metaphorical framework to reflect the complex and multidimensional nature of evolution as described by niche construction theory (Odling-Smee et al., 1996). Instead of a chess game, imagine a basketball court representing an ecosystem—a niche where humanity and other species interact according to certain rules. These rules, akin to the laws of nature, dictate that all living and non-living entities must adapt and respond to their environment.

This basketball court is not static; it evolves. Over time, it transforms into different playing fields—a soccer pitch, a volleyball court, even a virtual gaming space—each with its own set of rules. This continuous transformation highlights that survival and success do not depend on being the tallest, the strongest, or the most skilled in a specific game but on the ability to adapt to changing environments.

Consider the scenario where the basketball court, symbolizing a stable ecosystem, splits into smaller units, each governed by new rules as time progresses. This metaphor emphasizes that evolution is not about linear progress but about flexibility and adaptability. Success in this ever-changing game requires an intrinsic capacity to adapt, a trait shaped by the cumulative experiences and adaptations of previous generations.

Thus, evolution emerges as a multilayered interaction among animate and inanimate agents, each influencing and being influenced by their environment. This perspective underscores that survival is not about the perfection of any single player but the ability to navigate and thrive within a continuously evolving landscape.

Gould's "The Panda's Thumb" (2010) describes the panda's thumb as an imperfect yet functional adaptation, demonstrating how evolution is less about achieving an ideal form and more about finding workable solutions within specific environmental contexts.

The evolving nature of this game reveals a collective, multidimensional, and multigenerational adaptability, which can be described as a form of creativity at the biological level that better explains the evolutionary engine than mere competition. Just as the panda's unique thumb allows it to thrive in bamboo forests, this creative adaptability enables organisms to flourish in diverse environments, portraying evolution as an allegorical "performative art," a raceless dance—a series of adaptive movements and responses that transcend rigid frameworks.

Through Gould's insights, we can appreciate that the creative adaptations we observe in nature are akin to artistic performances, where each species contributes to a larger, ever-changing composition.

This perspective encourages to move beyond conventional notions of progress and embrace a more holistic understanding of evolution, where adaptability and creativity are celebrated as essential components of life's ongoing dance.

Niche Construction and Collective Creativity: Art, Agency, and Environmental Transformation

This section of the chapter will emphasize that the term "creativity" is not merely a metaphor borrowed from the artistic realm to be applied in biology. Instead, it is a fundamental structural feature essential for a comprehensive understanding of biological processes.

Although creativity is often celebrated as a unique individual talent, it is fundamentally a manifestation of associative thinking—a universal human ability that surfaced approximately 150,000–200,000 years ago. The true essence of creativity shines when it exerts a collective influence on a specific environment. Theoretical frameworks suggest that this environment can be examined through cultural and anthropological, social and political, or physical and abstract perspectives. However, an evolutionary biology perspective on creativity is seldom explored.

To do so, the concept of niche construction, introduced in the previous section, could aptly help describe the relationship between human creative agency and environmental modification, which, in turn, affects the relationship of other humans and, potentially, non-humans inhabiting that specific environment.

Niche construction refers to the process by which an organism modifies its local environment, whether its own or that of another species (Odling-Smee et al., 1996). These modifications can involve physical changes to the environment or the organism moving to a different habitat to experience new environmental conditions. Examples include animals building nests and burrows, and plants creating shade, altering wind speeds, and changing nutrient cycles. While these changes often benefit the organism making them, this is not always the case. For instance, when organisms discard waste, they can inadvertently degrade their own habitats.

The impact of an organism's actions is not limited to isolated instances but can lead to extensive environmental changes. A prime example is the behavior of beavers. Beavers build dams, which alter the flow of rivers and create ponds. This activity significantly transforms the landscape, affecting not only the water flow but also the surrounding ecosystem. The creation of ponds can lead to the development of new habitats for various species, increased biodiversity, and changes in vegetation patterns.

The beavers' dam-building behavior is a common trait and is repeatedly manifested by many individuals within the species. This collective activity results in substantial and lasting environmental modifications. Thus, the beavers' niche construction demonstrates how an organism's consistent and widespread behavior can reshape its environment, highlighting the broader impact of such ecological interactions.

Similarly, creativity becomes significant when art is produced multiple times by different individuals, amplifying its collective impact. Artistic agency gains importance when these creations are repeated, though not identically, evolving into collective efforts that influence many. These impacts transcend mere expressions of individual talent or progress.

Niche construction, reminiscent of the idea of evolution as a dance, not a race, highlights the reciprocal relationship between living organisms and their inanimate environment. Organisms modify their habitats, and these transformed environments, in turn, influence the organisms themselves. In this way, art, as a form of environmental modification, shapes how future generations will continue to alter that environment. For example, the Renaissance mastery of perspective radically transformed cities from the fifteenth century onward. Although the Renaissance is often viewed as a departure from the Middle Ages, successive generations of artists reshaped medieval cities into new ecosystems, setting the stage for this cultural rebirth and its lasting impact on urban landscapes. Through the lens of non-linear progress, neither epoch represents a perfect or inevitable result of artistic evolution; rather, reality is but one expression of many possible outcomes (Pievani, 2023).

The reciprocal and essential relationship between the cultural and political transformations of the Middle Ages and the Renaissance revolution does not imply any scale of progress. The Renaissance, in evolutionary terms, represents just one of many potential transformations emerging from that ecosystem, and not necessarily the best.

Furthermore, a transformation might introduce features that, in unforeseen ways, eventually become burdens rather than benefits. In this evolutionary context, while a single painting might seem revolutionary, its true significance lies in the features that gather the most extensive and diverse influences from within and beyond the realm of art.

These features make the that specific painting particularly meaningful within its specific context, society, or a particular style or technique. However, they also determine a course of evolution with all its positive and potentially negative impacts, which can manifest only under certain conditions.

The heuristic value of a concept like niche construction underscores that the impact of creative agency is not limited to a singular occurrence. Instead, it gains significance through repeated manifestations by various individuals, thereby amplifying its transformative power. Artistic agency becomes particularly notable when creations are reiterated—though not identically—evolving into collective endeavors that make a significant beneficial impact on a broader audience. In this way, it transcends individual talent or progress.

Viewing agency in the environment as a form of performative art is bolstered by the exaptive perspective, which posits that organisms do not necessarily modify their surroundings with a specific functional intent. Many actions lack a defined purpose, yet their success depends on their potential for functional cooptation at certain historical moments. This concept of exaptation (Gould & Vrba, 1982), particularly through the crucial notion of spandrels—derived from both a structural and artistic component of St. Mark's Cathedral in Venice (Gould & Lewontin, 1979)—is inherently linked to the idea of creativity, art, and performance.

For niche construction to influence evolution, three "performative" conditions must be met, irrespective of the function attributed to the transformation:

1. The organism must significantly alter environmental conditions.
2. These alterations must impact one or more selection pressures on a recipient organism.
3. There must be an evolutionary response in at least one recipient population due to the environmental modification.

Meeting the first two criteria alone provides evidence of niche construction. The same applies to art performance. Here, the organism assumes the role of the performative artist, the recipient becomes the audience, and the population represents the wider community. In performative arts, the audience and artist might be interchangeable, just as in extended taxonomies, these roles can interchange.

Therefore, art, as an evolutionary byproduct, is not evaluated by its functionality or immediate progress but serves as a medium for transforming societal conditions. Categories like beauty function as feedback mechanisms to assess creativity beyond its perceived purpose. In line with exaptation, this creativity thrives on redundancy, variability, and diversity. The beauty of an art performance, therefore, represents a promise, or even a wager, on the possibility of a positive transformation, as its success can be determined by functional cooptations that do not necessarily apply immediately or directly. Consequently, creative potential increases as collective and environmental conditions become more complex, diverse, and variable. By analogy, greater genetic diversity and favorable environmental conditions foster more biodiversity and the emergence of unexpected and "creative" organisms.

Squirrels Cannot Find 80% of the Nuts. This is How Trees Are Planted

In the arts, a larger and more diverse society with favorable conditions for artistic production is likely to fuel more varied, unexpected, and potentially radical and revolutionary forms of art. Just as biodiversity can rapidly change and even lead to mass extinctions, the conditions for artistic production can also shift dramatically. This perspective helps explain the existence of creative expressions like the construction of Göbekli Tepe and potentially other manifestations that we have overlooked or have yet to discover. It confirms that the complexity of the arts results more from collective and environmental input than from individual creativity alone. However, our understanding of history often leaves little room to explain why Göbekli Tepe was a more refined expression of art than many subsequent manifestations, as well as the fact that this settlement was eventually abandoned and lost to memory.

Continuing with the analogy to biology, considerations on the axis of creativity moving towards collective efforts rather than individual ones raise questions about the current understanding of the phylogeny and ontogeny of the arts. The phylogeny of the arts, rather than depicting the development and diversification of a species or group of organisms, illustrates the history of the evolution of the arts through various epochs and styles. Similarly, the ontogeny of art, which showcases how an individual artist's expressions evolve from embryonic ideas to mature works throughout their life, mirrors the development of an individual organism or a specific feature of an organism. Through these notions, evolution seeks to explain why something exists. However, once taxonomies, categorizations, and classifications are established, they tend to crystallize, making it difficult to discern the nuances among these categories and leading to the loss of pieces of the narrative. For a certain period, the relationship between ontogeny and phylogeny in biology was misunderstood but considered factual, which also led to dramatic societal transformations (Gould, 1985).

Let's consider the current concept of the progress of the arts as a phylogenetic tree, where multidisciplinary correlations with the arts are organized chronologically. For instance, a portion of this tree spans from the construction of Gothic cathedrals to the invention of perspective, both of which encompass expressions of performing arts. However, true cross-disciplinarity is often elusive because the historical development of art forms has already branched out and crystallized into distinct categories. Integrating multidisciplinarity within this context requires constructing abstract, artificial bridges between these branches. In doing so, we may overlook more natural and organic relationships that existed within the original framework.

Once the crystallization of categories occurs, discussions on art, even scholarly ones, shift from evolutionary notions to historiographical and critical perspectives, applying the values of cultural anthropology to the aforementioned classifications. These discussions might also include ethical positions, which are absent in evolutionary biology. The term "performative arts" suggests a common ground with evolutionary mechanisms, the evolution of architecture, cities, and territories can also belong to this milieu, both in terms of process and as the stage of life at the core of the performance. However, once the evolutionary perspective of art history shifts to ethics and then to a moral compass, as in the case of classical mimetic theories discussed later in the book, it becomes an obstacle to understanding the evolution of art beyond the moral crystallization of existing categorizations.

Only more recently have studies on performative arts, and the arts in general, shifted from the idea of imitation of reality to exploring their origins and evolutionary nature. In this sense, biomusicology has been seminal. However, even within this field, the analysis has lagged behind biological evolution research, with the evolutionary nature primarily examined through an adaptationist lens.

The crystallization of an ethnographic approach encompasses the finalistic perspective of art as the imitation of nature. For instance, primitive forms of performing arts, such as cave drawings and dance, are interpreted as imitations of nature, simulating hunting gestures or animal movements to reinforce a sense of community belonging.

Studies such as "Origins of Music" (Wallin, 1991) and research on the birth of creativity (Pringle, 2013) focus on afinalistic evolutionary mechanisms, posing different and intriguing questions. The ability to enhance community resilience is, therefore, a consequential functional cooptation of the inception of music and dance, which likely derive from a transformation of parts of the brain linked to taste. Performative art can thus be considered a spandrel.

Why do we engage in music and dance? Probably because, from an evolutionary perspective, we are equipped with the ability to facilitate serendipitous associations from our creative drive, enabling us to be both exuberant in our actions (performance) and in the things we produce (fine arts). Being creative here is not about making something for a specific use but about being exuberant in the sense of increasing redundancy, variability, and diversity from the physical features of human habitat. Thanks to associative thinking, the results of these processes can then be used for various purposes, even though they were not originally created/ designed for them. When this happens, there is likely a persistence of evolutionary successful traits through a complex system of feedback. We are somehow pure artists who prefer to see ourselves as engineers. And when we use these terms here, we are referring to the current interpretation, as the initial part of the book might also be considered a way to stress that engineers are also artists.

If we apply this view to reflect on traditional art literature, we might find it insufficient that a historical period like the Baroque is often analyzed exclusively as an aesthetic and formalistic response to a changing society, a reflection of that society, or a mannerist evolution or devolution from a significant architectural period like the Renaissance. From a conventional perspective—more aligned with an orthodox adaptationist evolutionary view-, the significance of this art form lies in whether it meets a functional need for which the art itself was designed and evolved, such as social cohesion, empathy, communication, or language.

From an exaptive perspective, a particular style or artistic structure endures because it is a byproduct of our inherent creative traits. These traits persist because even what we deem as pure formalism or decoration may fulfill certain structural, social, or cultural needs at a given moment, albeit in ways not easily comprehended. In this context, cognition can be envisioned as a semi-graph of this latent potential, even when the practical application of a certain architectural component, which we find aesthetically pleasing, has yet to be realized. Jumping forward a few centuries, we might see in Loos's "Ornament and Crime" (1929) the epitome of a centuries-long transformation, where a crystallized limited taxonomy evolves into a moral compass.

References

Austin, J. L. (1962). *How to do things with words.* Oxford University Press.

Butler, J. (1990). *Gender trouble: Feminism and the subversion of identity.* Routledge.

Childe, V. G. (1936). *Man makes himself*. Watts & Company.

Goldberg, R. (1979). Performance art: From futurism to the present. Harry N. Abrams.

Gould, S. J. (1996). *Full house: The spread of excellence from Plato to Darwin.* Harvard University Press.

Gould, S. J. (2010). *The Panda's thumb: More reflections in natural history.* WW Norton & Company.

Gould, S. J. (1985). *Ontogeny and phylogeny.* Harvard University Press.

Gould, S. J., & Lewontin, R. C. (1979). The spandrels of San Marco and the Panglossian paradigm: A critique of the adaptationist programme. *Proceedings of the Royal Society of London. Series B. Biological Sciences, 205*(1161), 581–598.

Gould, S. J., & Vrba, E. S. (1982). Exaptation—A missing term in the science of form. *Paleobiology, 8*(1), 4–15.

Graeber, D., & Wengrow, D. (2021). *The dawn of everything.* Farrar.

Loos, A. (1929–2019). Ornament and crime. Penguin

Melis, A., & Melis, G. L. (1996). *Architettura pisana.* ETS

Melis, A., Pievani, T., & Lara-Hernandez, A. J. (2024). *Architectural exaptation.* Routledge.

Odling-Smee, F. J., Laland, K. N., & Feldman, M. W. (1996). Niche construction. *The American Naturalist, 147*(4), 641–648.

Pievani, T. (2023). *Imperfection: A natural history.* MIT Press.

Pringle, H. (2013). The birth of creativity. *Scientific American.*

Rudofsky, B. (1987). *Architecture without architects: A short introduction to non-pedigreed architecture.* UNM Press.

Wallin, N. L. (1991). *Biomusicology: Neurophysiological, neuropsychological, and evolutionary perspectives on the origins and purposes of music* (Vol. 68). Pendragon Press.

Chapter 2
From Creativity to Evolutivity

Alessandro Melis and Marco Pelle

Abstract This chapter focuses on creativity as a mechanism mirroring evolutionary processes, proposing that creativity fosters connections and adaptations beyond individual expression. By likening artistic endeavors to evolutionary biology concepts such as Punctuated Equilibrium, the chapter posits that creativity reflects episodic leaps in human culture and cognition, enabling rapid adaptations amid stability. This view suggests that art possesses an epigenetic influence, extending its impact across generations by shaping behavioral and cognitive evolution. Introducing evolutionary concepts like exaptation and niche construction, the chapter reinterprets art as a manifestation of 'involutivity'—an ability to associatively assign values and purposes to performance and objects rather than creating ex nihilo This perspective challenges conventional definitions, emphasizing creativity as a dynamic interplay of adaptive and exaptive processes, shaping cultural taxonomy with polymathic influences. Through case studies spanning various historical periods, the chapter illustrates how art interweaves with scientific, philosophical, and technological paradigms, highlighting its transformative potential within society. Ultimately, redefining creativity as "involutivity" underscores the capacity of the arts to foster diverse pathways and collective resilience, mirroring evolutionary mechanisms that expand human adaptability through associative thinking, variability, and the creative reuse of forms.

Keywords Creativity · Evolutionary processes · Connections and adaptations · Artistic endeavors · Punctuated equilibrium · Human culture · Cognitive evolution · Epigenetic influence · Generational impact · Exaptation · Niche construction · Involutivity · Associative values · Adaptive processes · Cultural taxonomy · Polymathic influences · Case studies · Scientific paradigms · Philosophical paradigms · Transformative potential · Collective resilience · Associative thinking · Creative reuse · Variability · Interdisciplinary connections

A. Melis (✉)
School of Architecture and Design, New York Institute of Technology, New York, USA
e-mail: amelis@nyit.edu

M. Pelle
Department of Social and Political Sciences, Bocconi University, Milano, Italy

A. Melis and M. Pelle (eds.), *Design for Performative Arts Spaces*,
https://doi.org/10.1007/978-3-031-98215-6_2

Creativity has the power to extend experiences beyond the individual sphere. When considering epic literature, rock 'n' roll, or cinema, this expansion reaches even beyond one's immediate community and time, making dissemination and impact more effective within the broader human community.

Creativity thus serves as an evolutionary mechanism, multiplying the potential for associations beyond sexual and genetic means. Indeed, art possesses epigenetic potential, influencing not only individual development but also cultural and societal evolution over generations. This hypothesis, which suggests that artistic expressions and experiences can shape and transform our genetic expressions and behavioral patterns, will be further investigated in the next pages, exploring its implications on human creativity and cultural inheritance.

Understanding creativity as an evolutionary concept necessitates the introduction of further principles from evolutionary biology, in addition to what was already discussed in Chap. 1. This enables a deeper insight into contemporary theories that have challenged traditional notions, such as the gradual nature of evolutionary mechanisms. By incorporating these revolutionary biological principles, we can explore how performing creativity may have evolved in more complex and dynamic ways, reflecting not just slow, incremental changes, but also rapid, transformative leaps that have significantly influenced human development and cultural progression.

The evolutionary notion of Punctuated Equilibrium, for instance, describes a pattern of evolution in which long periods of stasis (where species remain unchanged) are punctuated by shorter periods of rapid change during which new species arise quickly from a common ancestor. This theory was proposed by paleontologists Stephen Jay Gould and Niles Eldredge in 1993 to explain the patterns they observed in the fossil record, which often showed abrupt changes in form and structure within species over relatively short geological periods, rather than gradual evolutionary changes (Gould & Eldredge, 1993). This perspective reinforces the consideration of performative art as a scaled reflection of evolutionary mechanisms, including dynamic change of rhythms and choreographies involving agencies and environmental pressures, as described in Chap. 1.

Heuristically, a punctuated equilibrium within society and culture can describe art as oscillating between stable and revolutionary forces, enriching both traditional classifications and avant-garde movements. Within a traditional society, art fulfills a taxonomical role distinct from science. However, in the avant-garde periods of a more mobile society, the boundaries between art and science become increasingly blurred. This convergence and alignment lead to a fusion where artistic and scientific classifications intertwine and influence each other, favoring polymathic perspectives.

Performative arts span across time and cultures, offering a mirror to society's values, struggles, and aspirations. The origins, meanings, and varied expressions of performative arts across different cultures can help interpret the rhythm between long periods of stasis and revolutionary tensions. The examination of case studies in the following part of this book, supported by a range of scholarly literature, aims to provide a comprehensive understanding of the performative arts' role in shaping and reflecting human experience within changing societies.

How Science and Society Shape Creative Expression

While investigating the realm of performative arts, this book aims to explore the interconnections among disciplines through cultural taxonomies and their mutual influences. For instance, it explores the reciprocal impacts of philosophy and performative arts in Ancient Greece, the unbreakable link between mysticism and art during the Middle Ages, and how creationism and humanism became essential for Renaissance art. Additionally, it examines the influence of evolutionism, sociology, anthropology, psychoanalysis, and technology on the concept of art. These disciplines have not only reshaped our understanding of art but have also been instrumental in defining distinct educational perspectives across various disciplinary areas.

Throughout history, art has alternated between anticipating scientific knowledge and mirroring existing knowledge through artistic expression. At times, it has supported traditional categorizations, reinforcing societal inertia in religion, politics, and science. Within these societies, the categorization of art as a separate and autonomous discipline might itself be an expression of a particular societal need, while multidisciplinarity reflects a more mobile societal structure. In other historical contexts, art can even be seen as an exploratory component embedded within other forms of knowledge, beyond multidisciplinarity, as exemplified by the Renaissance polymath.

Today, new discoveries in biology and neuroscience could be essential for developing a more nuanced view of art taxonomy that acknowledges its evolutionary complexity. Separating creativity from other brain mechanisms poses a risk, as a simplified understanding of art may resist more complex and articulated perspectives.

Among these risks, the gap between a discovery and the incorporation of an idea into educational programs may result in texts with outdated scientific content, making society resistant to change. It will take time for this more comprehensive understanding to spread and become ingrained in societal values, as observed with the Copernican Revolution, psychoanalysis, and other paradigm shifts.

Even today, despite overcoming creationism, Darwinian orthodox evolutionary positions still make many believe that a chimpanzee is a primitive version of a human, that we are at the apex of a linear evolution, that evolution revolves around the survival of the strongest, and that binary societal roles reflect nature. Furthermore, the belief that the Homo sapiens male brain is the largest among humans and directly correlates, for instance, with superior intelligence compared to women persists. Even subtle remnants of recapitulation theories support forms of unconscious racism.

The extension of the evolutionary taxonomy towards exaptation, as described by Elizabeth Vrba and Stephen Jay Gould in their influential work "Exaptation—A Missing Term in the Science of Form" published in 1982, or considering the concept of punctuated equilibrium (1982), highlights how the resistance of societies, accustomed to traditional forms of evolutionism, to change perspectives is likely evolutionarily consistent. These concepts challenge the gradualist view that evolution occurs at a steady, almost imperceptible rate and suggest instead that significant evolutionary changes can happen in relatively short bursts, followed by long periods of stability.

This extension and refinement of Darwinian theory reflect the complexity of evolutionary mechanisms and underscore the need for educational curricula to evolve in light of new scientific discoveries to provide a more comprehensive understanding of biological evolution.

Pievani, for instance, in "The Unexpected Life" (2011) introduces concepts such as serendipity in evolution, emphasizing the role of chance and the non-linearity of evolutionary processes. This perspective complements and expands upon Darwin's theories by addressing more recent discoveries in evolutionary biology and genetics. It suggests that evolution is not solely driven by natural selection but is also significantly influenced by random events and environmental factors. These insights underline the necessity for educational programs to continually adapt and integrate the latest scientific developments to ensure their content remains relevant and accurate.

If perceived solely as a reflection of societal values and autonomous as a discipline, or multidisciplinary only in the sense of including soft science knowledge, making it impermeable to revolutionary changes and expansions in other fields of knowledge, art might lose its potential to explore novel notions, to be at the forefront of research, a catalyst of change, and a means of providing feedback to science and society. In the case of performing art, this limitation will involve both content and processes as part of the evolutionary mechanism.

Limiting art to imitating nature and representing emotions and ideas may cause it to lose the ability to express new concepts if they are not yet widely known. A new scientific perspective could propose that art goes beyond simply representing reality or interpreting ideas; it also serves as an evolutionary mechanism that fosters creativity in various forms.

Emerging knowledge may instead encourage novel methods of thinking and interacting with the environment, necessitating new categorizations and terminologies that connect art to broader concepts. Such an approach helps in perceiving art holistically, acknowledging its links to something beyond traditional classifications.

Simplifications that prevent from grasping the complexity and correlations of evolution are always lurking around the corner... In a scene from the recent television adaptation of the novel "The Three-Body Problem" (2024), the aliens suggest that if our ancestors had not experienced fear, they might have gone extinct. This implies that bravery, displayed by some individuals, is considered a risk. It raises the question: why does humanity sometimes honor the fearless as heroes? Although this dilemma is yet unresolved in the series, this scene highlights the simplistic dichotomy of rational versus emotional responses, where the value of individualistic heroism is preserved as a traditional ethos.

In another segment of the series, an alien remarks that humans resemble insects once they realize their own limitations. Yet, is it accurate to given for granted that humans superior to insects? From an evolutionary perspective, insects have demonstrated greater success than humans, raising the question of which species is closer to extinction. This challenges the notion that insects are of lesser value and suggests that the misjudgment of human intelligence stems from a subjective and perhaps arrogant viewpoint (Gould, 1996). The aliens' statements mirror human classifications, rooted in a traditional biological outlook, and seem to adopt a viewpoint akin to that of an orthodox evolutionist. Meanwhile, human perspectives on how to overcome extreme crises appear overly simplistic, reiterating the epic narrative of mythological heroes.

The challenges to our survival are manifold and complex: relying solely on a single response tailored to a specific threat poses a risk to our survival, given the vast array of possible scenarios. The lesson from biology is to include a greater diversity of responses to both expected and unexpected risks, rather than focusing solely on whether fear or its absence is the correct response to a risk scenario. Individuals exhibiting fear, those without it, and every nuance within the fear spectrum are considered in response to a multitude of threats.

This redundancy of possibilities is an evolutionary mechanism that generates more than is needed for threats that might not happen. It can address both predictable and unpredictable events, meaning that some evolutionary paths will succeed, others may lead to dead ends, while some remain unexplored. The persistence of these exuberant, maximalist traits throughout history shows that redundancy is an acceptable cost, challenging the minimalist, rationalist view of evolution as producing only what is needed for an expected scenario.

Translating this to creativity, we should favor expanded categorizations with a spectrum of nuances where technology, art, and science intersect in ways unlike traditional disciplines, which often view engineering as the opposite of art.

Evolutionary mechanisms can be both deterministic, designed to address expected scenarios, and artistic, characterized by purposelessness and feedback mechanisms related to pleasure and values such as aesthetics, which cannot be fully understood through rational awareness. Some authors have likened this complexity to tinkering. However, bricolage may be seen as a reductionist metaphor for creative impulse. This suggests the need for an extended categorization, potentially requiring new terminology—perhaps "evolutivity"?

If Art Is Not A Manifestation of Creativity (Ex Nihilo), but Instead A Manifestation of Evolutionary Mechanisms, Should We Call It Evolutivity Instead of Creativity?

The word "creativity" is an example of how a concept can become inertial, even when scientific knowledge renders it obsolete and contradictory. "Creativity" represents an idea in which humans, seen as the pinnacle of evolution, possess a distinctive intelligence allowing them to create something ex nihilo, embodying unique metaphysical potential. The term "creativity" originates from the Latin verb "creare," which means "to create, to make." Its initial meaning emphasized the act of bringing something into existence that did not previously exist—especially from nothing, a concept often attributed to divine creation.

The journey of the word "creativity" from its Latin origins to its current usage encapsulates the transition from a concrete action ("to make") to an abstract quality or ability encompassing originality, inventiveness, and imagination. Initially, the concept of creation was closely tied to divine or mystical processes. During the Renaissance, it was translated into humanistic terms, meaning that creativity, once a prerogative of God, became a prerogative of mankind, seen as central to the universe. However, over time, especially with the Enlightenment's emphasis on reason and the Romantic era's focus on individual expression, the concept broadened to include human capacities for innovation and artistic expression.

In the twentieth century, "creativity" began to be studied scientifically in neuroscience and through the lens of psychoanalysis, leading to a further democratization of the term: it came to be understood not just as a capacity for artistic invention but as a fundamental human ability to solve problems, think critically, and generate novel solutions across all areas of life.

Throughout history, as the meaning of "creativity" expanded, we observe a shift in agency—from gods, to selected individuals, and eventually to all humans—while the notion of creating from scratch remained a constant. Humans, whether as reflections of divinity or as protagonists, have always been at the center of this action. Even today, human creativity is widely recognized as evidence of our place at the pinnacle of progress. This is often depicted in popular iconography as a line of transformation, with Homo sapiens at one end and a monkey at the other, reflecting an alleged unique sense of consciousness that distinguishes humans from other beings. In the Western tradition, shifts in agency are particularly evident in art. For example, our ability to imitate nature has been viewed in different ways: either as divine inspiration, reflecting the Christian mystical perspective of the Middle Ages, or as an assertion of human divine origin, reflecting the humanistic view of empowerment to create.

Although in contemporary secular society art, which is not directly influenced by religious or scientific values, is still seen as something created ex nihilo—attributed uniquely to humans and hence termed "creation" and "artificial"—new scientific discoveries are leading to a more complex and nuanced understanding. We have learned that multiple human species coexisted on Earth, that our cognitive abilities may be less unique than previously thought (Gould, 1996), that brain size is less significant than once believed, and that we are not the only species capable of creating art.

The expansion of evolutionary biology taxonomies, including the introduction of exaptation and the sequencing of the human genome, has introduced more uncertainties than certainties regarding the meaning of creativity, which has been largely overlooked by art literature. We now know more, yet we also understand that many of our earlier beliefs were mistaken. For instance, exaptation describes evolution as a mechanism that favors the functional co-option of existing structures. Imperfection and redundancy of shapes can transform existing objects into spandrels, which serve as reservoirs of possibilities (Gould & Lewontin, 1979). However, society tends to resist radical changes. When we refer to "junk" DNA, we reveal our struggle to comprehend evolutionary disorder, which disrupts the notion of a world governed by orderly laws. In 2013, Heather Pringle reviewed new paleoanthropological studies and found that creativity arises from the development of neuronal interconnections, rather than being solely a function of brain size. This process involves generating thoughts and actions through associative thinking, rather than creation ex nihilo.

Thus, creativity is recognized as stemming from an alternative or complementary approach to linear thinking. It involves building new relationships among existing ideas and introducing new uses for existing forms, rather than creating something entirely new. On the contrary, biological studies on morphospace (Melis et al., 2024) suggest that the possible remains much larger than the real, and that nature—and humans as a byproduct of nature—still do not utilize all the possible shapes that could be potentially effective.

This semantic shift redirects the discussion of creativity from seeing it merely as the ability to create something that did not previously exist to a deeper understanding grounded in neurology and psychology, emphasizing a perspective that is less human-centric. "Creativity" might be more accurately described as a product of "evolution," reflecting an impulse fueled by the development of associative thinking—an enhancement that occurred as our brain's neuronal connectivity increased, rather than its size—to manipulate shapes, repurpose objects and materials, and associate new uses with existing forms. This perspective presents at least three unconventional consequences: (1) the novelty of creation lies in the use of shapes or objects, not the objects themselves, which can be purposeless; (2) function follows form, rather than vice versa; (3) there is no inherent distinction a priori in the creative process for art, technology, and science.

The three points implicitly suggest that creativity encompasses evolving ideas and practices in ways that may not be immediately obvious or intended. As a byproduct of evolution, mostly based on the functional cooptation of shapes and the rejection of reverse engineering—since function does not necessarily correspond to its original use, and shape is not dictated solely by function (Melis et al., 2024)—it supports the hypothesis that nature is more of an artist than an engineer. Rational processes merge with afinalistic methods in a combinatory manner, challenging the term "creativity" when examined etymologically.

Rather than creating from nothing, "evolutivity," or any other alternative term to creativity which embeds the heuristic values of biology, could express the ability to assign new, multiple, or alternative uses—including meanings—to existing structures that are either unused, meaningless to previous humans, or have outdated uses. This also includes building objects available in the morphospace, meaning that the potential for these shapes already exists in reality (Melis et al., 2024), and assigning functions to them. Evolutivity simulates, at the human level, evolutionary mechanisms through a combination of adaptive and exaptive actions at a strategic level. In this context, the goal is not to invent shapes that don't exist, but to increase, through the manipulation of materials, redundancy, variability, and diversity of shapes which can therefore become vessels of possibilities within expected or unexpected scenarios. Uses and functions are these possibilities.

References

Gould, S. J., & Lewontin, R. C. (1979). The spandrels of San Marco and the Panglossian paradigm: A critique of the adaptationist programme. *Proceedings of the Royal Society of London. Series B. Biological Sciences, 205*(1161), 581–598.

Gould, S. J. (1996). Mismeasure of man. WW Norton

Gould, S. J., & Eldredge, N. (1993). Punctuated equilibrium comes of age. *Nature, 366*(6452), 223–227.

Gould, S. J., & Vrba, E. S. (1982). Exaptation—A missing term in the science of form. *Paleobiology, 8*(1), 4–15.

Melis, A., Pievani, T., & Lara-Hernandez, J. A. (2024). *Architectural exaptation: when function follows form.* Taylor & Francis.

Pievani, T. (2011). *La Vita inaspettata.* Raffaello Cortina Editore.

Pringle, H. (2013). The birth of creativity. *Scientific American.*

Chapter 3
Origins of Performance

The Blurring Confines of Art Categories

Alessandro Melis and Marco Pelle

Abstract This chapter explores the origins of performative arts, focusing on their unique alignment with "evolutivity"— a concept that sees creativity as an adaptive-exaptive, associative mechanism beyond linear categorization. Performative arts, from theater and dance to spoken word, embody live experiences where the act of performance itself is central, reflecting life's transient nature and engaging audiences in dynamic interactions. Unlike cinema, which captures performances for repeated viewing, performative arts rely on the immediacy of presence, highlighting their flexibility and adaptability. This chapter also examines prehistoric and cross-cultural origins of performance, including ritual burials, early musical instruments, and petroglyphs, as holistic expressions of early human creativity. By integrating evolutionary biology, it contends that such forms of expression arose as adaptive social practices, fostering community bonds and cultural transmission. These arts are positioned as essential to early human survival, serving as social adhesives and channels for transmitting knowledge and cultural identity. Tracing performative arts to their roots reveals their role in human development as reservoirs of possibility, transcending survival functions. The chapter ultimately suggests that these early practices, rather than evolutionary byproducts, were fundamental to cultural progression, interweaving performative and visual arts into a unified, evolutionary mechanism.

Keywords Performative arts · Evolutivity · Creativity · Adaptive-exaptive mechanism · Associative mechanism · Linear categorization · Theater and dance · Spoken word · Live experiences · Transient nature · Audience engagement · Immediacy of presence · Flexibility and adaptability · Prehistoric origins · Cross-cultural performance · Ritual burials · Early musical instruments · Petroglyphs · Holistic expression · Evolutionary biology · Adaptive social practices · Community bonds · Cultural transmission · Social adhesives ·

A. Melis (✉)
School of Architecture and Design, New York Institute of Technology, New York, USA
e-mail: amelis@nyit.edu

M. Pelle
Department of Social and Political Sciences, Bocconi University, Milano, Italy

A. Melis and M. Pelle (eds.), *Design for Performative Arts Spaces*,
https://doi.org/10.1007/978-3-031-98215-6_3

Knowledge transfer · Cultural identity · Human development · Reservoirs of possibility · Cultural progression · Unified evolutionary mechanism

Performative arts are perhaps the most consistent with the concept of "evolutivity." These arts involve live performances before an audience, making the act of performance integral to the art itself. By replicating reality holistically, they incorporate the performers' agency, rather than focusing solely on the final product.

Examples include theater, dance, music, opera, performance art, and spoken word activities like poetry readings and stand-up comedy. These forms represent and embody life through their temporal nature, existing in the moment and often relying on performer-audience interaction.

Performative arts can be presented in various venues—from traditional theaters and concert halls to unconventional spaces like streets or galleries. This flexibility and adaptability enhance their alignment with evolutionary mechanisms and the concept of "evolutivity," which eludes simple categorization within specialized or monodisciplinary frameworks.

Cinema exemplifies the challenges inherent in classifying entities that exist on a spectrum rather than within distinct, separate categories. While cinema incorporates elements of performance, it is traditionally categorized separately from performative arts such as theater or dance. The primary distinction lies in the nature of the performance and its presentation.

Performative arts are defined by live performances in front of an audience, where the immediacy and interaction between performers and the audience play a crucial role. These performances are ephemeral, existing only in the moment, and can vary from one presentation to another.

Cinema, conversely, involves the recording of performances—be it acting, dancing, or other forms of expression—which are then edited and presented as a finished film. This process allows for multiple viewings of the same performance, preserving it in a way that live performances cannot. Thus, the audience engages with the work from a distance, not in real-time with the performers.

However, the boundaries between cinema and performative arts can become blurred, especially with the advent of experimental cinema and performance art films, where the act of filming itself becomes part of the live performance art. Additionally, live cinema events, interactive film experiences, and other innovative formats are continually expanding the boundaries of what cinema can encompass.

In a broader sense, cinema can be considered a form of art that includes performative elements. Yet, it is primarily classified under visual arts due to its reliance on filmed and edited content rather than on live performance. While this classification may appear contemporary, it is as old as Plato's concept of mimesis, the multilayered imitation of reality (where performative arts constitute a sort of first layer).

Diverging from the conventional linear perspective of evolution, which typically encompasses separate and historically subsequent streams, cinema instead demonstrates that creativity—as part of an evolutionary process involving exaptation—tends toward a form of cultural entropy. In this context, the increase in complex and redundant associations plays a crucial role, in contrast to alternative streamlined pathways.

The Emergence of Performance in Art

The current book is part of an extensive exploration aimed at understanding whether current categorizations in the field of creativity effectively capture its nature. The first installment in this series commenced with research on architectural exaptation, aiming to comprehend creativity through the heuristic lens of evolutionary biology (Melis et al., 2022).

Following this initial inquiry, the second phase of research investigated the emergence of the arts as integral to human activity, challenging the notion of art as a standalone or evolutionarily recent phenomenon. Instead, it presented art as a development that transcends the conventional context of architectural typologies for exhibition. It argued that the concept of art exhibition as distinct from other human needs, organized through a specific taxonomy, is a relatively recent development. This taxonomy, in some cases, has shown its limitations by artificially segregating human needs, which are inherently interconnected in a holistic manner.

This book aims at better elucidating the potential of creativity through performance as a reservoir of possibilities. It presents performance as an emergency mode of survival, an alternative to the standard modality of linear thinking, which is often associated with the idea that creativity is not a primary mechanism or need in times of crises or emergencies. Through practice-based research cases, this book focuses on the performative arts within a dual historical narrative.

On one hand, it acknowledges the widely accepted ethnographic interpretation that performative arts arose from specific societal and individual human needs, leading to the creation of structured forms such as theatre and musical performances.

On the other hand, it examines the evolution of performative arts as an intrinsic human trait present from the outset of our species, manifesting in various forms, including those associated with architectural typologies of performative arts. These typologies are considered specific categorizations of performative arts, but not necessarily the only, most important, or most advanced ones.

Evolutionary Timeline and Cognitive Development

The narrative of human evolution is intimately linked to the development of performative arts, tracing back to key moments in the evolutionary timeline of Homo sapiens between 200,000 and 150,000 years ago. During this period, significant advancements in the human brain enabled the emergence of complex behaviors and cognitive skills essential for the creation and appreciation of performative arts. These evolutionary developments likely provided the necessary conditions for the emergence of language, symbolic thought, and the capacity for imaginative expression and representation.

Performative arts, from ritual dances to storytelling around the campfire, played crucial roles beyond mere entertainment—where, through exaptive interpretation, entertainment is a feedback mechanism. They acted as social adhesives, strengthening community bonds, transmitting knowledge, and supporting group cohesion and survival. Although, as mentioned in the initial chapters of the book, there is no evidence that this functionality is the cause of performative art or whether, conversely, performative art is a spandrel and these functionalities are functional co-options, in both hypotheses, evidence shows the importance of these evolutionary milestones.

Abstract Thinking and Metaphorical Representation and Paleoanthropological Insights on The Origins of Performative Arts

Performative expression, influenced by evolutionary pressures favoring cooperation, social learning, and emotional engagement and empathy, reflects humans' capacity for abstract thinking and the representation of complex concepts through metaphor and allegory. This ability allowed early humans to explore beyond their immediate experiences, exploring spiritual and metaphysical domains and paving new paths for artistic and cultural expression.

Through paleoanthropology, we gain insights into how performative arts have served as a foundation for human creativity, offering a means for our ancestors to navigate and articulate the depth of human experience well before the advent of written language or formal art institutions.

The origins of performative arts can be traced back to early human societies, where rituals, ceremonies, and storytelling were integral to communal life. These early forms of performance served not only as entertainment but also as means of transmitting cultural values, histories, and social norms. Over centuries, these practices evolved into more structured forms of performative arts, reflecting the cultural, political, and social dynamics of their times (Schechner, 2003).

Recent findings suggest that the capacity for artistic expression and symbolic thought was not unique to our species. Neanderthals, in particular, have been shown to engage in behaviors that suggest a form of artistic expression. For example, in Spain, cave paintings have been dated to more than 64,000 years ago—20,000 years before the arrival of Homo sapiens in Europe. These paintings include geometric patterns, hand stencils, and possibly symbolic markings, suggesting that Neanderthals were capable of abstract thought and symbolic expression.

Although evidence of aesthetic behavior or symbolic expression in Homo erectus is speculative and controversial, some researchers point to the Acheulean handaxes, crafted by Homo erectus over a million years ago, as having an aesthetic component due to their symmetry and form, which went beyond functional necessity. Similarly, the intentional disposal of dead bodies within the Dinaledi Chamber, if confirmed, may suggest a level of cognitive complexity and perhaps a rudimentary form of symbolic expression in Homo naledi.

Identifying art from hominin species predating Homo sapiens is fraught with challenges. Firstly, the perishable nature of many materials means that potentially artistic creations made from wood, skin, or plant matter have not survived. Secondly, the interpretation of what constitutes art is heavily influenced by modern human perspectives, making it difficult to assess the intentions behind ancient artifacts. Lastly, the absence of a clear context for many findings makes it hard to distinguish between functional objects and those that might have had symbolic or aesthetic significance.

Performative Arts and The Non-deterministic Nature of Prehistoric Categories

The notion that art predates Homo sapiens challenges the belief that our categorizations are universal and not simply products of more recent societal constructs rooted in specific historical contexts. Take, for instance, the discovery of fire. Evidence indicates that early humans, or their predecessors, mastered the use of fire several hundred thousand years ago. This achievement, which predates the advent of cave paintings, signifies a profound evolution in human cognitive and cultural capabilities.

Importantly, there is no evidence of determinism, often considered the backbone of human discoveries. The discovery of fire can be viewed as a form of performance. It is unlikely that our ancestors actively sought fire with a specific practical use in mind. Rather, it seems more plausible that the utilization of fire was a serendipitous adaptation to something already encountered or accidentally discovered. The act of engaging with and manipulating fire—its mastery, the fascination it sparked, and perhaps the rituals that arose around it—can indeed be seen as a form of performance.

This raises the question: did this purely artistic action persist until fire was associated with practical uses, such as cooking and heating? Is an activity considered a performance only if it lacks a practical application? By "practical," do we refer to activities that fulfill basic survival needs, such as eating and drinking? This perspective might be too restrictive, as the concept of performance could extend beyond mere survival necessities to include any activity that involves ritual or aesthetic pleasure.

The act of associating with and manipulating fire highlights the importance of viewing technological advancements not merely as milestones of human ingenuity but as deeply intertwined with the cultural, social, and possibly artistic practices of early humans. This hypothesis is supported by evidence of other early human species engaging in behaviors that might be considered cultural or ritualistic, indicating a complex social life involving shared practices and knowledge transmission.

From another perspective, the question of whether this constitutes art is complex. Historically, 'art' has been associated with actions linked to aesthetic pleasure rather than practical use, which would categorize it as technology. However, paleoanthropological or neurological perspective expand creativity to include technology, highlighting the thin or arbitrary distinction (Pringle, 2013).

The association of the discovery of fire with forms of performative arts, particularly as collective and ritual activities, presents a fascinating hypothesis that intersects the fields of archaeology, anthropology, and art history. This perspective considers the discovery and utilization of fire not merely as a technological milestone but as a cultural and possibly performative act embedded within early human societies. It opens intriguing avenues for understanding the complexity of early human societies and underscores the need for more comprehensive taxonomies in understanding human creativity and agency.

Burial as an Embryonic Form of Performing Art

Early human burials, with their complex rituals and symbolic components, are early expressions of performative art. These burials were not just about disposing of bodies practically—a modern abstraction—but were deeply tied to societal values, beliefs, and aesthetics. Examining grave goods, intentional arrangements, and ritualistic elements reveals how early humans communicated and perpetuated cultural identities. Through these burial practices, our ancestors used performance and aesthetic expression to navigate death, marking these rites as embryonic forms of performing art.

Unlike speculative discussions about performances around the fire, burial rites offer clearer insights into early societal structures, beliefs, and artistic expressions. The earliest confirmed intentional burials, dating back 100,000–120,000 years, at sites like Qafzeh and Skhul in the Near East, demonstrate early humans' complex social behaviors and cognitive abilities to conceive of life beyond death or express profound respect for the deceased.

For example, the Sungir burials in Russia, about 28,000 years old, showcase elaborate grave goods like thousands of ivory beads, indicating significant investment of time and resources. These artifacts, alongside specific arrangements of bodies and items, convey messages about social status, community bonds, and beliefs regarding the afterlife.

Burials as performative art extend beyond practical objectives. Rituals such as the careful placement of bodies, inclusion of grave goods, and use of materials like ochre were ways for early humans to communicate beliefs, values, and emotions. The performative aspect involved intentional actions and ceremonies observed by the community, creating a shared, communal experience. The aesthetic value in certain burials, evident in the arrangement of bodies and artifacts, suggests a concern with visual presentation and symbolic meaning, aligning with principles of aesthetic expression and collective performance.

Dinaledi Chamber: Ongoing Debates and Interpretations

The Dinaledi Chamber presents a captivating intersection of archaeology and paleoanthropology, shedding light on ancient hominin behaviors. Located in the Rising Star Cave system near Johannesburg, South Africa, the chamber gained scientific attention in 2013 when Lee Berger's research team announced the discovery of Homo naledi, a previously unidentified hominin species. This discovery has ignited a vibrant debate over evolutionary narratives and social behaviors (Berger, 2013).

The Dinaledi Chamber is a remote section of the cave, containing remains of at least 15 individuals. The intentional deposition of these bodies over time has led to speculation about Homo naledi engaging in ritualistic behaviors or deliberate disposal of the dead—actions once thought exclusive to Homo sapiens and Neanderthals.

The implications of the Dinaledi Chamber extend beyond academic interest. The inferred behaviors—whether or not they qualify as performative art or ritual—offer insights into early hominin adaptive strategies and potential symbolic interactions. Homo naledi, with its mix of archaic and relatively modern traits, challenges traditional evolutionary models and suggests a more nuanced evolutionary landscape.

The hypothesis of burial-like practices has sparked debate. Skeptics, citing the absence of direct burial evidence such as grave goods, suggest natural processes could explain the remains' accumulation. However, according to Berger, the chamber's inaccessibility and lack of alternative explanations support the argument for intentional disposal, implying significant insights into ancient hominin social structures and cognitive abilities.

Validating these behaviors would enhance our understanding of ancient hominin cognitive and social capabilities, indicating symbolic thought, cultural expressions, and an early conceptualization of life and death within hominin branches previously not credited with such complexity.

Performative Interpretation of Cave Paintings

Cave paintings are among the earliest forms of human artistic expression. The Chauvet Cave in France, dating back approximately 32,000–30,000 years ago, is a prime example. Researchers like Jean Clottes, who has extensively studied these caves, suggest that the deep, often inaccessible locations of these paintings imply a ritualistic or sacred intention behind their creation (Clottes, 1996). Similarly, the Altamira Cave in Spain and the Sulawesi caves in Indonesia provide further evidence of early human artistry, with paintings that range from 35,000 to 40,000 years old (Aubert et al., 2014).

Expanding the interpretation of cave paintings to include performative art enriches our understanding of these ancient works and their creators. It allows us to consider the paintings as part of a larger cultural and ritualistic practice that involved not just the visual senses but also movement, sound, and storytelling. This holistic view opens up new avenues for exploring how early humans understood their world and expressed their beliefs, fears, hopes, and stories through a combination of art and performance. By recognizing the performative aspects of cave paintings, we gain insights into the complex social and spiritual lives of our ancestors.

George Bataille's inquiries have been seminal for this discussion, which he articulated in his work, notably in "The Cradle of Humanity: Prehistoric Art and Culture" (originally "Lascaux ou la Naissance de l'Art," 1955). Bataille saw cave paintings not just as early forms of art but as expressions of something deeply fundamental and sacred to human nature. His interpretation explored themes of eroticism, death, and the sacred, suggesting that these artworks are manifestations of humanity's earliest attempts to confront and understand its existence and mortality.

For Bataille, the act of creating art in these ancient times was inherently linked to the sacred and the transgressive. He believed that the creation of art involved a violation of the natural order, a stepping beyond the boundaries of everyday life into the realm of the sacred or the taboo. This view aligns with his broader philosophical inquiries into the limits of experience and the interplay between taboo, transgression, and the sacred.

Bataille's interest in eroticism as a force that drives much of human behavior and thought is well-documented. He saw eroticism as intimately connected with death and the sacred, a theme that he extended to his interpretation of prehistoric cave art. In the depiction of animals and human figures, Bataille perceived an expression of humanity's deep-seated fascination with and fear of death, as well as a celebration of life and fertility. Thus, for Bataille, cave paintings could be seen as early performative expressions of the human attempt to reconcile with the inevitability of death and the continuation of life.

One of the most significant examples Bataille discussed was the Lascaux Cave in France. He considered Lascaux not merely as an archaeological site but as a sacred space where prehistoric humans engaged with profound existential questions. The animals depicted on the walls were not just subjects of hunting; they were also totems and symbols of deeper spiritual significance, embodying the sacredness of life and the mystery of existence.

Reinterpreting cave paintings as performative art, alongside their recognition as visual art, significantly broadens our understanding of these ancient creations and their societal roles. This perspective views the paintings not merely as static images but as dynamic elements in rituals or communal activities involving dance, music, storytelling, or enactments.

One compelling interpretation is that creating these paintings was a ritualistic performance. Many cave paintings are located deep within caves and hard-to-reach places, suggesting they were not intended for casual viewing. This implies a selective audience or possibly no human audience at all, indicating a more spiritual or communal purpose. The paintings might have been integral to ceremonies or rituals, serving as a backdrop or focal point for performances that connected participants with the spiritual world, their ancestors, or the animals depicted on the walls.

Cave paintings could also have been part of storytelling or mythmaking processes, serving as visual aids or memory devices for storytellers to convey tales of the hunt, creation myths, or ancestral stories. The act of painting and the subsequent use of these images in storytelling can be seen as performative acts that engaged the community and transmitted cultural knowledge and values across generations.

Some researchers, such as David Lewis-Williams, propose that cave paintings are linked to shamanistic practices and altered states of consciousness (Lewis-Williams, 2002). From this perspective, the creation of the art itself might have been a performance enacted by shamans during trance states. The images could represent visions experienced by the shamans, serving as a bridge between the physical world and the spiritual realm. The performance aspect comes into play as the shaman interacts with these images during rituals, guiding the community through spiritual experiences.

Recent studies on the acoustics of caves have added another layer to the performative interpretation of cave art. Some caves with paintings have been found to possess exceptional acoustic properties, suggesting that sound (e.g., music, chanting, clapping) might have played a role in the rituals associated with these paintings. The creation and enjoyment of the art could have been a multi-sensory experience, incorporating visual, auditory, and possibly even kinetic elements as participants moved through these acoustically vibrant spaces.

Sculptures and Performance: Identities and Holistic Art

The exploration of performative aspects in the context of prehistoric art, such as the crafting and use of sculptures and figurines, underscores a probable shift from the more holistic and nuanced understandings of early human societies to the more categorized perceptions of modern times. This shift mirrors broader changes in how societies organize, categorize, and understand the world around them. Recognizing the potential for a more integrated approach to art in early human societies opens new avenues for understanding the complexity and depth of prehistoric social structures and their expressive practices. It reminds us that the divisions among art forms are largely a modern construct that may not accurately reflect the intricate nature of human creativity and expression throughout history.

In contemporary discourse, art is often divided into distinct categories such as visual art, performative art, and literary art. However, this compartmentalization overlooks the holistic nature of art in early human societies. For them, the creation of a figurine like the Venus of Willendorf might not have been seen purely as a sculptural or visual artistic endeavor but as an act imbued with multiple layers of meaning, encompassing both the performative and the spiritual.

While the Venus of Willendorf, estimated to have been created between 28,000 and 25,000 BCE, provides evidence of early human engagement with sculptural forms, scholars, including Christopher S. Henshilwood, suggest that such figurines extended beyond mere artistic endeavors, playing pivotal roles in prehistoric social structures, potentially in fertility rituals or social bonding practices (Henshilwood, 2009). This indicates an early recognition of art's power to influence, shape, and reflect societal norms and values.

The creation and use of these figurines could have been performances in themselves, rituals that engaged communities in shared experiences that transcended simple visual appreciation. They were objects of art that carried profound communal significance, possibly used in ceremonies or as totems that connected the community through shared beliefs and practices.

The early human perspective on art likely reflected a more integrated view of identity, where the distinctions between artist, artifact, and audience were less pronounced. Artifacts like the Venus of Willendorf were not just representations of human or divine forms but active participants in the social and spiritual life of the community. This holistic approach suggests that early humans recognized the fluidity and interconnectedness of identities, roles, and representations within their societies.

Petroglyphs as a Performative Stage

Performance art and petroglyphs, situated between the primitive forms of sculpture, painting, and writing, are linked by the concept of the stone as a stage for performance. Petroglyphs are images or symbols carved, etched, or engraved into rock surfaces by prehistoric peoples. The word "petroglyph" comes from the Greek words "petros," meaning "stone," and "glyphein," meaning "to carve." These carvings often hold significant cultural and historical importance, serving as a performative form of ancient communication. Petroglyphs provide insights into the lives, beliefs, and environments of prehistoric peoples. They are found worldwide and can depict various subjects, including animals, humans, abstract designs, and scenes from daily life or spiritual beliefs.

The Bhimbetka rock shelters in India, with petroglyphs dating back approximately 30,000 years, are significant for their early artistic and potential narrative content, indicative of early human attempts at storytelling, often linked with performative elements such as ritualistic practices (Dubey-Pathak, 2014). The rock carvings at Murujuga (also known as the Burrup Peninsula) in Western Australia are among the world's largest collections of petroglyphs, with images dating back at least 30,000 years. These carvings include marine life, human figures, and abstract designs. In Brazil, the petroglyphs of the Serra da Capivara National Park are believed to date back at least 25,000 years. These images include human figures, animals, and abstract shapes, offering insight into the region's prehistoric cultures (Guidon & Delibrias, 1986).

Rock paintings also hold significant importance, with one of the earliest known examples found in the cave of Lascaux, dating back to around 17,000 years ago. The petroglyphs at the Petroglyph National Monument in New Mexico, USA, are among the most well-known in North America. These carvings, created by Native American and Spanish settlers over a span of 3000 years, depict a variety of symbols, animals, and human figures (Schaafsma, 1986). The Valcamonica rock carvings in Italy are one of the largest collections of prehistoric petroglyphs in the world, with over 140,000 symbols and figures carved into rock faces, dating from the Epipaleolithic (8000–5000 BCE) to the Middle Ages. The petroglyphs at the Tassili n'Ajjer in Algeria are renowned for their depiction of prehistoric life in the Sahara Desert. These carvings, created between 6000 BCE and 1000 CE, show animals, human figures, and daily activities. The petroglyphs of the Altai Mountains in Russia are notable for their extensive range, with images dating back to the Bronze Age (approximately 3000–1000 BCE). They include depictions of animals, humans, and hunting scenes (Jacobson, 1993).

All these examples of petroglyphs contribute to a form of visual art that transcends time, allowing modern humans to connect with ancient cultures through the images left behind on rock surfaces. As such, petroglyphs remain a crucial subject of study in archaeology and anthropology.

Music and Dance in The Framework of Evolutionary Biology and Epistemology

According to UNESCO's definition, music is perhaps the most universal of the performing arts, present in every society. It is often an integral part of other performing art forms and domains of intangible cultural heritage, such as rituals, festive events, and oral traditions. Music can be found in diverse contexts, whether sacred or secular, classical or popular, and is closely linked to work or entertainment. It also carries political and economic dimensions: recounting a community's history, praising powerful individuals, and playing a role in economic transactions. Music is performed on various occasions, including marriages, funerals, rituals, initiations, festivities, and many other social functions (UNESCO, n.d.).

The exploration of music's role extends beyond cultural and anthropological contexts to the realms of evolutionary biology and epistemology. Biomusicology, a field that examines the biological underpinnings of music, offers one perspective that posits music as an adaptation shaped by natural selection, providing evolutionary advantages to our ancestors. Theories supporting music as an adaptation highlight its role in social cohesion, mate selection, and communication. For instance, Charles Darwin, in "The Descent of Man," speculated that musical ability might have been a sexual selection trait, helping individuals attract mates through displays of skill and creativity (Darwin, 1871).

The discovery of flutes made from bird bones and mammoth ivory in Germany's Swabian Jura region, estimated to be around 40,000 years old, points to the presence of music in prehistoric times (Conard et al., 2009). This date aligns with the birth of creativity (Pringle, 2013), corroborating the perspective that music played a crucial role in the creative and social evolution of early humans. Steven Mithen, in "The Singing Neanderthals," proposes that music and language co-evolved as a single communicative system. Mithen argues that early hominins used a "musilanguage," a precursor to both music and language, which facilitated social bonding and coordination within groups. This theory supports the idea that music contributed to the survival and reproductive success of early humans (Mithen, 2007).

Contrastingly, the concept of music as a spandrel, or exaptation, suggests that music emerged as a byproduct of other evolutionary processes. Stephen Jay Gould and Richard Lewontin introduced the term spandrel to describe features that arise as incidental byproducts of other adaptations. In this view, music did not evolve directly due to its own adaptive value but rather as a side effect of other cognitive and neurological developments, such as language and complex auditory processing (Gould & Lewontin, 1979).

Steven Pinker, in "How the Mind Works," famously referred to music as "auditory cheesecake," implying that music exploits pre-existing neural circuits that evolved for other purposes, like language and emotional expression. This analogy underscores the idea that while music may not have been directly selected for its own sake, it nonetheless provides significant pleasure and emotional engagement, capitalizing on cognitive faculties shaped by other adaptive functions (Pinker, 1997).

Biomusicology bridges these perspectives by exploring how music perception and production are rooted in the brain and body. Scholars like Nils L. Wallin, Björn Merker, and Steven Brown, in their edited volume "The Origins of Music," discuss how musical traits may have evolved and the neural mechanisms involved in musical behavior. Their work integrates findings from neuroscience, psychology, and evolutionary biology to provide a comprehensive understanding of music's biological basis (Wallin et al., 1999).

Even the etymology of the word "music," tracing back to the Greek word "mousike," which referred to the arts governed by the Muses, might reflect its evolutionary transformation from an exaptive trait into a narrative that responds to human satisfaction, transforming a certain level of complexity into a mythological recognition. This complexity often necessitates referring to metaphysical entities when the intricacies of evolution do not allow for an alignment between intentionality and rational use, with pleasure and functionality acting on an evolutionary timeline (Melis & Pievani, 2022; Wallin et al., 1999).

Both music and architecture involve the organization of elements to create structures that evoke emotional and aesthetic responses. Music, with its temporal structures of rhythm, melody, and harmony, can shape our perception of time and space, much like how architecture shapes our physical environment through rhythm and form of composition which humans perceive as harmonic. Architectural spaces can influence the properties of music, such as acoustics, while music can enhance the experience of architectural spaces. For example, the design of cathedrals embraces diverse forms of perception to enhance sensorial experience, including acoustics. Conversely, music can transform the ambiance of a space, affecting how it is perceived and used.

Dance, within this context, considering the ritual significance of movement in early humans, is a dynamic and ephemeral art form that bridges music and architectural space. This intricate relationship manifests through the way dance translates musical rhythms and emotions into physical movements that navigate and define the spaces in which they occur.

At its core, dance is an embodied response to music, where each gesture, step, and leap is a physical interpretation of the auditory experience. Music provides the temporal structure, setting the pace and mood, while dance interprets these elements through movement, creating a visual representation of the sound. The rhythm, melody, and dynamics of music are echoed in the dancer's body, making the invisible elements of music tangible and visible.

Architectural space, on the other hand, provides the physical context for dance. It is within this spatial framework that dancers move, interact, and create patterns. The architecture of a space—whether it's a grand theater, an open plaza, or an intimate studio—shapes and is shaped by the dance that takes place within it. Dancers use the dimensions, textures, and forms of the space to inform their movements, navigating around obstacles, utilizing heights and depths, and responding to the spatial constraints and opportunities.

Through dance, architectural spaces are redefined and brought to life. The interplay of movement and space creates a dialogue where the dancer not only occupies space but actively transforms it. Each movement traces an invisible line that temporarily reconfigures the spatial dynamics, creating a fluid architecture that exists only in the moment of the dance. This transient architecture is constructed through the dancer's interaction with the physical environment, highlighting and altering our perception of space.

Furthermore, the choreography of a dance can be seen as a form of spatial composition. Just as an architect designs buildings with an understanding of human scale and movement, a choreographer designs dances with an understanding of how bodies move through space. This spatial choreography considers the flow of movement, the use of levels, and the relationship between dancers and their surroundings. It is a form of spatial storytelling, where the narrative is conveyed through the interaction of bodies and space.

The connection between dance, music, and architectural space is a powerful one, as it demonstrates the unity of the arts in shaping human experience. Dance, as a physical manifestation of music, transforms and animates architectural space, creating a holistic sensory experience that engages sight, sound, and movement. This triadic relationship enriches our understanding of how space can be experienced and interpreted, not just as a static environment but as a dynamic and living entity shaped by human interaction.

In conclusion, dance serves as a crucial linkage between music and architectural space, redefining our perception of both. Through the physical embodiment of music and the dynamic interaction with architectural space, dance creates a unique and ephemeral form of art that transcends the boundaries of each individual medium. This interconnectedness highlights the transformative power of dance and its ability to bring together the auditory, visual, and spatial dimensions of human experience.

References

Aubert, M., Brumm, A., Ramli, M., Sutikna, T., Saptomo, E. W., Hakim, B., Morwood, M. J., van den Bergh, G. D., Kinsley, L., & Dosseto, A. (2014). Pleistocene cave art from Sulawesi Indonesia. *Nature, 514*(7521), 223–227.

Berger, L. (2013). *Homo naledi and the chamber of secrets*. National Geographic.

Clottes, J. (1996). Thematic changes in upper Palaeolithic art: A view from the Grotte Chauvet. *Antiquity, 70*(268), 276–288.

Conard, N. J., Malina, M., & Münzel, S. C. (2009). New flutes document the earliest musical tradition in southwestern Germany. *Nature, 460*(7256), 737–740.

Darwin, C. (1871). *The descent of man, and selection in relation to sex*. John Murray.

Dubey-Pathak, M. (2014). The rock art of the Bhimbetka area in India. *Adoranten, 5*.

Gould, S. J., & Lewontin, R. C. (1979). The spandrels of San Marco and the Panglossian paradigm: A critique of the adaptationist programme. *Proceedings of the Royal Society of London. Series B. Biological Sciences, 205*(1161), 581–598.

Guidon, N., & Delibrias, G. (1986). Carbon-14 dates point to man in the Americas 32,000 years ago. *Nature, 321*(6072), 769–771.

Henshilwood, C. S. (2009). Symbolism and the origins of modern culture: The emergence of symbolic culture recorded in African middle stone age artefacts. *South African Archaeological Bulletin, 64*(189), 121–129.

Jacobson, E. (1993). *The deer goddess of ancient Siberia: A study in the ecology of belief*. E.J. Brill.

Lewis-Williams, D. (2002). *The mind in the cave: Consciousness and the origins of art*. Thames & Hudson.

Melis, A., Lara-Hernandez, J. A., & Melis, B. (2022). Learning from the biology of evolution: Exaptation as a design strategy for future cities. *Smart and Sustainable Built Environment, 11*(2), 205–216.

Melis, A., & Pievani, T. (2022). Exaptation as a design strategy for resilient communities. *Transdisciplinarity* (pp. 307–327). Springer International Publishing.

Mithen, S. (2007). *The singing Neanderthals: The origins of music, language, mind, and body*. Harvard University Press.

Pinker, S. (1997). *How the mind works*. W. W. Norton & Company.

Pringle, H. (2013). The birth of creativity. *Scientific American*.

Schaafsma, P. (1986). *Indian rock art of the southwest*. University of New Mexico Press.

Schechner, R. (2003). *Performance theory*. Routledge.

UNESCO. (n.d.). Performing arts. Retrieved from https://ich.unesco.org/en/performing-arts-00054

Wallin, N. L., Merker, B., & Brown, S. (Eds.). (1999). *The origins of music*. MIT Press.

Chapter 4
Architectural Space for Performative Arts

Mimesis as The foundation of Classic Performative Arts

Alessandro Melis and Marco Pelle

Abstract This chapter explores the architectural evolution of spaces designed for performative arts, grounded in the principle of mimesis introduced by Plato and Aristotle. As a creative process capturing universal truths through imitation, mimesis has profoundly influenced performance spaces, from ancient Greek amphitheaters to modern digital environments. These spaces serve as more than functional backdrops; they actively shape and are shaped by the performances they host, embodying cultural shifts, technological advancements, and evolving societal values. The chapter traces the development of performance venues from the early amphitheaters of Greece and Rome, through the elaborate opera houses of the Baroque era, to the adaptive spaces of the twenty-first century. This progression reveals how physical design amplifies the audience's engagement and enhances the thematic depth of performances, transforming spaces into integral components of artistic expression. Examining performance spaces across different cultural and historical contexts, Chapter 4 highlights their role in fostering community, reflecting ideals, and enabling a spectrum of performative experiences. Through case studies, the chapter illustrates how the architectural space for performative arts is not merely a container but an evolving partner in the artistic experience, continuously responding to and shaping the dynamics of creativity, audience interaction, and artistic intent.

Keywords Architectural evolution · Performative arts · Mimesis · Plato and Aristotle · Creative process · Universal truths · Performance spaces · Ancient Greek amphitheaters · Modern digital environments · Cultural shifts · Technological advancements · Societal values · Opera houses · Baroque era · Adaptive spaces · Audience engagement · Thematic depth · Artistic expression · Cultural contexts · Community fostering · Ideals reflection · Performative experiences · Case studies · Evolving partner · Dynamics of creativity · Audience

A. Melis (✉)
School of Architecture and Design, New York Institute of Technology, New York, USA
e-mail: amelis@nyit.edu

M. Pelle
Department of Social and Political Sciences, Bocconi University, Milano, Italy

A. Melis and M. Pelle (eds.), *Design for Performative Arts Spaces*,
https://doi.org/10.1007/978-3-031-98215-6_4

interaction · Artistic intent · Integral components · Physical design · Performance venue development

The genesis of performing arts, rooted in Aristotle's "Poetics" and "Metaphysics," is anchored in the foundational theory of mimesis, which has permeated various epochs and thinkers, continually molding the trajectory of the arts (Adam, 1907; Aristotle, 1933). Aristotle laid its groundwork in "Poetics" by defining art as an imitation of life. He argued that this imitation, or mimesis, is a natural inclination in humans, driven by an instinct for harmony and rhythm. Aristotle's mimesis is not simple mimicry but a creative process involving selection and representation, aiming to convey universal truths. For Aristotle, the highest forms of art are those that represent actions and events to evoke catharsis, a concept particularly relevant to drama and theater.

Aristotle's "Metaphysics" further deepens this exploration by examining the nature of reality itself. Here, he discusses the principles of being, essence, and substance, providing a framework for understanding the world that artists seek to imitate. If we consider "Metaphysics" in relation to "Poetics," the act of imitation in art becomes a means to express underlying truths about existence and the nature of things beyond their mere appearances.

The connection between the imitative nature of art and the reality that is not directly perceivable invites a groundbreaking interpretation. If reality itself is a complex, layered construct that we cannot fully grasp, then the art of imitation (mimesis) becomes even more significant as an attempt to capture and convey aspects of this elusive reality, making it accessible to human perception and understanding.

Performing arts as an "imitation of imitation" are among second-order reflections on human action and the nature of reality. Performers mimic the actions, emotions, and narratives of life, which are themselves interpretations of the world. Thus, performing arts become a medium through which artists and audiences engage with and reflect on the nature of existence, perception, and the essence of reality.

This "imitation of imitation" suggests a layered approach to understanding and experiencing the world, where performing arts serve as a mirror to both the actions of humans and the deeper, metaphysical truths underlying those actions. It positions performing arts as a vital medium for exploring, questioning, and communicating the complexities of human existence and the nature of the world.

Prior to Aristotle, Plato had a different take on mimesis. In "The Republic," Plato regarded art and imitation with skepticism, viewing them as shallow copies of the forms that constitute reality, twice removed from the truth. Despite his criticisms, Plato's discussions on art and imitation contributed significantly to the dialogue surrounding mimesis, setting the stage for Aristotle's more favorable interpretation.

Representing the spirit of their time and context, the Greek tragedians Aeschylus, Sophocles, and Euripides can be seen as early practitioners linking the performative arts to the theory of mimesis, even though they predated Aristotle's writings. Their works presented human actions and divine interventions in ways that reflected the complexities of human nature and the cosmos, aiming for catharsis in their audiences. These tragedians, through storytelling and dramatic representations, laid the groundwork for the performative arts as a medium for exploring profound human and metaphysical truths.

The art of the Renaissance is a prime example of the long-lasting influence of the mimetic theory. This period, marked by a revival of interest in classical antiquity, saw artists and thinkers like Leonardo da Vinci and Michelangelo emphasize the imitation of nature as the highest form of art, echoing Aristotle's views. Their masterpieces sought to capture the beauty and complexity of the natural world, embodying the principles of mimesis in their pursuit of artistic excellence.

In literature, Shakespeare's plays vividly demonstrated mimesis in their exploration of human nature, society, and the cosmos. His works brought to life the complexities of existence through the imitation of action, creating an array of characters and narratives that continue to resonate with audiences today. Shakespeare's ability to depict universal truths through particular stories underscored the enduring power of mimesis in the performative arts.

The Romantic era and subsequent modern movements introduced new interpretations of mimesis. Romantic artists and writers, such as William Wordsworth and John Keats, while not always directly engaging with Aristotelian mimesis, shared the belief in art's capacity to capture and reflect universal truths through the particular. Their works emphasized emotional depth and individual experience, expanding the scope of mimesis to include the personal and subjective.

In the twentieth century, Bertolt Brecht's concept of the "Verfremdungseffekt" (alienation effect) in theater challenged traditional notions of mimesis by deliberately distancing the audience from an empathetic response, aiming to incite critical reflection instead. Brecht's innovative approach highlighted the dynamic and evolving nature of mimesis, demonstrating how performative arts could be used to provoke thought and inspire social change.

Philosophers such as Theodor Adorno and Max Horkheimer in "Dialectic of Enlightenment" and Hans-Georg Gadamer in "Truth and Method" have also engaged with the concept of mimesis, examining its implications for aesthetics, ethics, and understanding. These works expand the concept of the representation of reality beyond mere imitation to include the transformative power of art in society and individual cognition. Adorno and Horkheimer critiqued the culture industry's manipulation of mimesis for commercial ends, while Gadamer explored the dialogical nature of art and its role in revealing truth.

Stages of Performing Expression

Performative arts manifest differently across cultures, shaped by historical, social, and political contexts. For instance, classical Indian dance forms such as Bharatanatyam embody religious storytelling and movements, while African dance is deeply connected to communal life and rituals. In the West, forms like opera and ballet reflect the evolution of European artistic and cultural traditions over centuries (Williams, 1993).

This diversity extends not only to the content but also to the spaces that host these performances. Architecture and urban planning can be considered part of performative arts. Buildings and cities serve as stages for performances, and their evolution, ideation, and realization within specific historical contexts are performances in themselves.

The history of performance space architecture is more than a chronicle of structural innovation; it reflects the shifting dynamics of society, culture, and art. Each era's architectural advancements reveal evolving notions of community, performance, and the role of art in public life. As we continue to build and reimagine spaces for performative arts, we carry forward a rich legacy of creativity and communal experience, affirming the theater's enduring place in human expression.

The following sections will explore the relationship between physical spaces designed for performative arts and the artistic expressions they host. By examining the architectural evolution of performance venues, from ancient amphitheaters to modern digital spaces, we can understand how these environments shape and are shaped by the performative acts they accommodate. Through case studies and a diverse range of literature, we aim to illuminate this symbiotic relationship, revealing how spaces contribute to the cultural and aesthetic dimensions of performative arts.

Performance spaces play a crucial role in the experience and interpretation of art forms. Architecture, in this context, goes beyond mere functionality; it becomes an active participant in the performance, influencing everything from acoustics to audience interaction. This chapter explores how the design and structure of performance spaces impact the performative arts and, conversely, how performative practices influence architectural innovation.

Exploring performative arts through architecture reveals deep connections between physical spaces and artistic expression. This chapter examines historical and contemporary performance venues, from Greek amphitheaters to modern digital environments, highlighting how design shapes and is shaped by performance. It underscores the vital role of thoughtful architectural design in enriching artistic communities and fostering cultural innovation.

The Role of Space in Performative Arts

Space is not merely a backdrop for performance but an active participant that interacts with performers and the audience. Its design, shape, size, acoustics, and material textures all influence the perception and impact of a performance and reflect shifts in artistic expressions, cultural values, and technological advancements (Bachelard, 1964), from the acoustic harmony of ancient Greece's open-air amphitheaters to the exuberant machinery of Baroque opera houses and the flexible, immersive environments of contemporary venues.

The origins of performative art spaces can be traced back to ancient rituals and ceremonies with the first inklings of performance emerging from the religious practices of early civilizations.

The Egyptian Abydos Passion Play can be considered one of the earliest formalized expressions of theatre, performed around 2000 BCE, depicting the death and resurrection of the god Osiris.

Also in China, theatrical performances have been a part of court ceremonies since the Shang Dynasty (c. 1600–1046 BCE). However, it was during the Tang Dynasty (618–907 CE) that Chinese theater began to evolve into more sophisticated forms, including the Pear Garden, which is considered the first royal acting and music academy.

The inception of classical theater architecture can be traced back to ancient Greece in the sixth century BCE, where it evolved from the Dionysian rituals. These festivals featured performances of dithyrambs, choral hymns sung and danced by a group of performers, which, over time, evolved into a more structured form of drama.

The transition from ritualistic performances to a structured theatrical form is attributed to Thespis, a figure often regarded as the first actor, around 534 BCE. Thespis is said to have introduced the concept of an individual actor stepping out of the chorus to assume individual roles, thus creating a dialogue between the chorus and this new entity, the actor. This innovation marked the birth of theater as a distinct art form, leading to the development of tragedy, comedy, and satyr plays, which were performed during the city's many festivals.

The theater quickly evolved, becoming more sophisticated in terms of structure, performance, and thematic exploration. The construction of purpose-built theaters, such as the Theatre of Dionysus in Athens nestled on the southern slope of the Acropolis, provided a physical space dedicated to this art form, designed to accommodate large audiences and to enhance the acoustics and visibility of the performances.

By the fifth century BCE, Greek theater had reached its zenith, with playwrights by Aeschylus, Sophocles, Euripides, and Aristophanes contributing timeless works that not only entertained but also provoked thought, explored ethical and philosophical dilemmas, and critiqued society. Their works laid the foundational principles of drama and theatrical performance that would influence Western theater for millennia to come.

Therefore, while performance and theatricality preceded them, the Greeks are credited with the invention of theater as a specialized typology as we understand it today, with tiered seating and open stages, engineered for optimal visibility and acoustics. This design philosophy catered to the practical aspects of performance while embodying a deeper societal conviction, as Greek theaters served not only as performance venues but also as civic spaces fostering communal engagement and reflecting the era's belief in theater as a pivotal element of civic life.

Following the Theatre of Dionysus, another remarkable example of engineering, acoustic quality, and classical Greek architectural aesthetics was the Theater of Epidaurus. Constructed in the fourth century BCE, it could seat up to 14,000 spectators and is still used today for performances, particularly during the annual Epidaurus Festival.

Located within the sanctuary of Apollo, the Delphi Theater, with its stunning views of the entire sanctuary and the valley below, underscored the integration of culture, nature, and spirituality in ancient Greek life, and hosted musical contests and dramatic performances during the Pythian Games, part of the four Panhellenic Games.

The relationship between theater architecture and performance was greatly influenced by Aristotle's "Poetics", leading to spaces that were not merely functional but deeply intertwined with the contemporaneous understanding of drama and performance art: Aristotle's analysis of drama as a structured form of storytelling that elicits catharsis underscores the importance of the audience's experiential journey, and the architectural design of Greek theaters—aimed at enhancing visibility and auditory clarity—directly complemented his principles by ensuring that performances effectively communicated narratives and evoked emotional responses from the audience.

The holistic view of physical spaces accommodating the changing dynamics of performance, serving not merely as venues but as active contributors to the performative arts, was inherited by the Romans. They added their own architectural innovations, most notably the use of the arch, which allowed for the construction of more ambitious structures that reflected the might and cultural aspirations of the Empire and hosted a variety of performances, from plays to gladiatorial contests.

The Theatre of Pompey, considered the first permanent Roman theater, was built in 55 BCE by Gnaeus Pompeius Magnus, and could hold about 20,000 spectators. It was a grand structure, notable for its lavish use of marble and for including a temple dedicated to Venus Victrix. Its design set a precedent for future Roman theaters and amphitheaters.

Another significant Roman theater, the Theatre of Marcellus was commissioned by Julius Caesar and completed by Augustus in 13 BCE. It could accommodate approximately 11,000 spectators and was praised for its harmonious proportions and the integration of Greek architectural elements. Parts of its structure have been preserved, and it remains a prominent landmark in modern Rome.

Located in what is now Turkey, the Aspendos Theatre is one of the best-preserved examples of Roman theater architecture. Built in the second century CE during the reign of Marcus Aurelius, it is renowned for its majestic façade and excellent state of preservation. The theater, which could seat over 15,000 spectators, is a testament to the Roman Empire's architectural influence across its territories.

Performative art underwent significant transformation from classical antiquity to the Middle Ages. While classical times saw it as a celebration of civic pride, entertainment, and a means to explore philosophical and existential questions, the rise of Christianity and feudal society shifted its focus to religious instruction and moral guidance, moving from celebrating human achievements and questioning the gods to reinforcing Christian values and communal cohesion. While classical theater focused on mythological tales, historical events, and the human condition, medieval theater, emerging prominently around the tenth century, was predominantly aimed at educating the populace about biblical stories and moral virtues.

This transition reflects broader changes in society, where the focus of community life shifted from the public sphere of the Greek agora and the Roman forum to the church and local festivities. Performative art became more localized and integrated into the rhythms of agricultural and religious life, marking a departure from the more centralized and state-sponsored productions of the classical world.

Overall, the evolution from classical to medieval theater highlights the adaptability of performative art to changing societal structures, values, and technologies. It demonstrates the enduring capacity of theater to reflect and shape human understanding of the world.

Unlike the fixed stone grand theaters of ancient Greece and Rome, medieval theater between the 5th and late fifteenth centuries lacked permanent architectural structures for performances and often utilized temporary structures or open public spaces, such as town squares. The production of medieval plays was typically a community effort, often sponsored by local guilds or the church, as opposed to the state-sponsored festivals of classical antiquity. The audience in medieval times was more diverse, including people from various social classes. The interactive nature of performances, such as processions and outdoor stages, allowed for a more engaged and participatory audience.

This period of transition highlights how medieval theater, adapting to the changing dynamics of society, moved from the monumental architectural settings and mythological themes of classical antiquity to more ephemeral structures and Christian-centric narratives, reflecting broader cultural and religious shifts.

Originating from within the church services, liturgical dramas were initially simple dramatizations of biblical stories, such as the Passion of Christ, gaining popularity from the tenth century onwards. They were performed in Latin and integrated into the church's liturgical services. Over time, some of these performances moved outside the church to reach a broader audience, marking the beginnings of a theater that was both part of and distinct from its religious origins.

Mystery Plays, which flourished from the Twelfth to the sixteenth centuries, were among the most popular forms of medieval theater, presenting stories from the Bible in a series of short scenes or plays. Often performed on movable wagons (pageant wagons) that traveled through towns, these plays were produced by local guilds. An example is the York Mystery Plays from England, with records indicating performances as early as the fourteenth century, which are still performed today.

Morality plays, emerging in the late medieval period around the fifteenth century, were allegorical dramas designed to teach moral lessons to their audiences. Characters often personified moral qualities (Virtue, Vice) or abstract concepts (Death, Mankind). One of the most famous morality plays is "Everyman," with the oldest surviving edition dated from the early sixteenth century, which explores themes of salvation, death, and redemption.

During the same period in Asia, long-lasting theatrical forms such as Indonesia's Wayang Kulit shadow puppet theater, dating back to at least the tenth century and still performed today, utilized carved leather puppets to tell stories from Hindu epics and local folklore. Similarly, Cambodia's Khmer shadow theater showcased classical dance dramas rooted in the Ramayana and local myths, reflecting the region's rich storytelling traditions through dance and music. Concurrently, the Yuan Dynasty (1271–1368 CE) in China saw the flourishing of Zaju, a drama form integrating poetry, music, and dance, while the subsequent Ming (1368–1644 CE) and Qing (1644–1912 CE) periods witnessed the rise of Peking Opera, a synthesis of music, vocal performance, mime, dance, and acrobatics. In Japan, during the Muromachi period (1336–1573), Noh theater, patronized by the warrior class, emerged as a refined, stylized performance art blending dance, drama, music, and poetry. Central to Noh's aesthetic are its masks, which convey a wide range of emotions and themes, transforming actors and inviting audiences into a world where tradition, spirituality, and human emotion converge.

The structure of Noh plays, categorized into five main themes—gods, warriors, women, madwomen, and demons—reflects a deep exploration of existential questions and human experience. This structure is supported by the unique stage design of Noh theaters, featuring a main stage (honbutai) and a connecting bridge (hashigakari), which serves both symbolic and functional purposes.

The bridge represents the path between the physical and spiritual worlds, enhancing the mystical atmosphere of the performances. The chorus and musicians, employing traditional instruments like the flute (nohkan) and drums (kotsuzumi and ootsuzumi), add to the contemplative mood, making Noh a holistic sensory experience.

The Theatrical Evolution from the Fifteenth to The Nineteenth Century

Theater played a pivotal role in the European Renaissance as well, with the Teatro Olimpico in Vicenza, designed by Andrea Palladio, standing as the first and oldest permanent indoor theater of the modern era. Commissioned by the Olympic Academy for the staging of classical comedies, its construction began in 1580, coinciding with Palladio's death, and was completed with its inauguration on March 3, 1585, showcasing the renowned fixed scenes crafted by Vincenzo Scamozzi.

The project was based on interpretations of Vitruvius's "De architectura" and enriched by firsthand examinations of Roman theater ruins still standing at the time, notably the Teatro Berga in Vicenza. Recent research indicates that Palladio originally envisioned a singular perspective aligned with the central door of the scene, intending for painted backdrops to adorn the lateral arches. Furthermore, the strategic cutting of the two wall wings and the design of the "ducal" ceiling above the proscenium are attributed to his innovative project.

Scamozzi contributed with adaptations and enhancements to fulfill Palladio's vision and with wooden scenes, renowned for their perspectival illusionism and intricate detail, specially created for the opening spectacle.

The classical vocation of the theater was confirmed by the grand opening on March 3, 1585, featuring Sophocles' "Oedipus Rex," accompanied by the choirs of

Andrea Gabrieli—a performance revived in 1997 by the Olympic Academy under Gianfranco De Bosio's direction. On this and subsequent rare occasions, the scenes depicting the seven streets of Thebes were brought to life with a sophisticated and original artificial lighting system conceived by Angelo Ingegneri, a Venetian who had joined the Olympic ranks by April 1580. Crafted from wood and stucco for what was meant to be a temporary installation, these scenes have, against all odds, survived to this day, remaining extraordinarily well-preserved despite the risks of fire and damage from wartime bombings.

England experienced its own theatrical renaissance in the late sixteenth and early seventeenth centuries. The Globe Theatre, constructed in 1599 in London, became emblematic of this era, renowned for staging Shakespeare's plays. Its open-air, circular design captured the spirit of Elizabethan theater, fostering a unique connection between the performers and their audience. This period saw a flourishing theatrical scene, with authors such as Christopher Marlowe, Ben Jonson, and Thomas Kyd exploring profound themes of power, identity, and humanity.

The architectural design of Elizabethan theaters, characterized by thrust stages surrounded by galleries, enabled intimate actor-audience interactions and versatile dramatic presentations. The absence of elaborate scenery required the plays' language to evoke vivid settings and emotions, demanding unprecedented levels of engagement from audiences. Moreover, these plays mirrored the era's social, political, and religious tensions, offering veiled critiques that turned theaters into arenas for public discourse and examination of the day's pressing issues.

Amidst this vibrant theatrical culture, Inigo Jones, in the seventeenth century, emerged as a pivotal figure, bridging the gap between traditional Elizabethan and modern Stuart theater through his innovative designs. As the first significant English architect with a profound impact on theatrical design, Jones introduced the proscenium arch to the English stage, drawing inspiration from his studies in Italy and the works of Renaissance masters. This feature, which frames the stage and separates the audience from the performance, transformed the staging and viewing of plays by incorporating movable scenery and machines for rapid scene changes.

Jones' collaboration with Ben Jonson on royal court masques, such as "The Masque of Blackness" and "The Masque of Queens," showcased his revolutionary use of perspective scenery and architectural elements within the performance space. These lavish productions at the Banqueting House in Whitehall exemplified the synergy between architecture and performance art, marking a significant evolution in theater design. Jones' legacy in theater architecture is distinguished by his focus on visual harmony, spatial organization, and the integration of new technologies, which laid the foundation for the future of theater production design.

Also in the early seventeenth century, Kabuki emerged in Japan as a more accessible and engaging spectacle, contrasting with Noh. Founded by Izumo no Okuni around 1603, Kabuki quickly gained popularity for its dramatic makeup, elaborate costumes, and sophisticated stage mechanisms. Unlike Noh, Kabuki features all-male casts, with actors specializing in male (tachiyaku), female (onnagata), and villainous (katakiyaku) roles. This tradition, stemming from various bans on female and young male performers, has become a defining characteristic of Kabuki.

Kabuki is well-known for its "mie" pose, which highlights dramatic moments, and the "hanamichi," a runway extending into the audience that facilitates dramatic entrances and exits. These elements, along with revolving stages and trap doors, contribute to Kabuki's dynamic and visually captivating performances. The repertoire of Kabuki is diverse, ranging from historical narratives to domestic dramas, often incorporating contemporary themes while maintaining traditional elements. This adaptability has ensured Kabuki's relevance and popularity through the centuries.

Both Noh and Kabuki have significantly influenced modern theater, cinema, fashion, and literature by emphasizing the synthesis of various art forms, combining visual, musical, and narrative elements to create a holistic approach that resonates with audiences worldwide.

Grandiosity and Rationalism from Baroque to Neoclassicism

During the Baroque Period, theaters and opera houses epitomized the era's grandeur, reflecting the divine and majestic through elaborate ornamentation and rich colors. The introduction of the proscenium arch stage was a significant innovation, framing the stage like a picture to enhance depth and perspective, while facilitating elaborate scenic designs and stage machinery for special effects. These features catered to the Baroque love of spectacle and were prominent in courtly performances, with many theaters built within palaces or as standalone structures financed by the nobility. Opera, in particular, was popular, and theaters of this period served the tastes of the elite, becoming architectural embodiments of the era's artistic values.

The Teatro San Carlo in Naples, Italy, constructed in 1737, is one of the oldest working theaters in the world, showcasing the Baroque blend of technical engineering advancements and aesthetic values, including elaborate decorations and an ornately frescoed ceiling. The Margravial Opera House in Bayreuth, Germany, dating back to 1748, and the Drottningholm Palace Theatre in Sweden, from 1766, both feature the characteristic proscenium arch stage of the era. These theaters were equipped with innovative machinery for special effects, highlighting the Baroque fascination with creating an immersive theatrical experience.

Royal patronage played a crucial role in the proliferation of opulent performance spaces, exemplified by the Versailles Royal Opera in France, inaugurated in 1770 for the court of Louis XVI, and the Theater of the Palace of Caserta in Italy. These venues were designed not just for performances but as symbols of monarchical power and prestige, hosting operas and ballets that reflected the sophisticated tastes of the aristocracy. The integration of theaters into royal palaces, adorned with mirrors, chandeliers, and sumptuous materials, was a hallmark of the period, reflecting a society where different standards of lifestyle and aesthetic values separated the noble classes from the common people.

The Enlightenment marked a departure from the ornate private opulence of the elitist Baroque era, favoring ideals of public common good and a commitment to rationalism.

Theaters of the Enlightenment embraced a more orthodox Neoclassicism, reflecting the period's fascination with the virtues of the Greek polis and its idealized values. Architecturally, this meant a more restrained use of classical orders and the abandonment of the ornate excesses of the waning Rococo style. Notable examples include the Burgtheater in Vienna, Austria, founded in 1741, which became a hub for intellectual debate and the dissemination of Enlightenment ideas through drama. Another influential example, which shaped the aesthetics of many subsequent theaters interpreted as Attic temples, is Piermarini's La Scala in Milan.

During this period, theater became more accessible to the general public, reflecting Enlightenment ideals of education, civic engagement, and the belief in art's potential to improve society. In particular, France's Revolutionary architects and thinkers transformed performance spaces into integral societal venues for entertainment, reform, and inspiration (Adorno et al., 2002).

The spread of encyclopedic knowledge played a crucial role in the democratization process. The "Encyclopédie," edited by Denis Diderot and Jean le Rond d'Alembert during the Enlightenment, provided comprehensive articles on various subjects, including architecture and theater, fostering a culture of critical thinking and innovation. Diderot's extensive writings on the dramatic arts, which promoted naturalism in performance and staging, significantly influenced contemporary and later theater design and direction.

Étienne-Louis Boullée's "Cenotaph for Newton" (1784), while not a theater, embodies principles of creating monumental spaces that evoke awe and contemplation, which are also relevant to performance spaces. This concept aligns with the intrinsic human need for the celebration of burial and memory, discussed in the previous chapter and reflected in Boullée's theoretical works.

Claude-Nicolas Ledoux took a more utopian and socially oriented approach, exemplified by the Royal Saltworks at Arc-et-Senans. He envisioned theaters and other public buildings as key components of his ideal city, where architecture would play a direct role in educating and elevating the public.

Architects like Jean-Nicolas-Louis Durand (1760–1834) emphasized the technical focus of the École Polytechnique, contrasting with the Beaux-Arts' approach to functionality in architecture. His systematic approach to design, particularly in "Précis of the Lectures on Architecture" (1802–1805), influenced the efficient and versatile planning of performance spaces through rational planning and modular coordination of space.

Romantic Emotion and Sublime

The Romantic era of theater aimed to integrate entertainment with moral and philosophical improvement, exploring and expressing the depths of human emotion and the complexities of the psyche (Gadamer, 2004). Emphasizing the sublime, Romantic theater sought to evoke profound emotional responses from audiences, contrasting with the rationalist outlook of the Enlightenment. Advances in lighting and scenographic setups allowed designers to manipulate mood and atmosphere more effectively, enhancing the theatrical experience. From a typological point of view, this era's performance spaces continued to consider classical architecture as the image of the theater outlook, although classicism began to blend with Gothic or exotic revival styles.

In his 1851's "The Four Elements of Architecture" (1851), Gottfried Semper articulated fundamental principles of architecture of the period, discussing designs that combined functionalism with ornamentation, embodying the ideals of historicism.

Gottfried Semper's most famous contribution to theater architecture is the Semperoper in Dresden, originally completed in 1841 and rebuilt after a fire in 1869 under the supervision of his son, Manfred Semper, due to his political exile. The 1878 Neo-Renaissance masterpiece blends functionality, acoustical excellence, and aesthetic beauty, reflecting Semper's belief that theater design should elevate public architecture to an art form that educates and inspires by emphasizing visual and spatial experiences alongside practical performance aspects.

His ideas about material honesty, structural clarity, and the symbolic representation of architecture had a lasting impact on the development of modern architectural theory and practice.

Semper's legacy in theater architecture is characterized by a balance between innovation and tradition, function and form, making him a pivotal figure in the evolution of performance space design.

The Bayreuth Festspielhaus in Germany, built by Richard Wagner in 1876, also exemplifies the Romantic spirit in its design, emphasizing an intimate connection between the performer and the audience, with innovative features like a hidden orchestra pit to enhance the acoustic experience and focus attention solely on the drama unfolding on stage.

Avant-Garde Movements and Modernism

A shift from architectural classicism in performance spaces occurred at the end of the nineteenth century with Art Nouveau, which replaced classical orders with organic, flowing lines, asymmetrical shapes, and details inspired by botany and other natural forms. An example is the Maison du Peuple, completed in 1899 in Brussels as the headquarters for the Belgian Workers' Party, designed by Victor Horta to accommodate various community functions, including meeting rooms, offices, a café, and multipurpose spaces, although it was not specifically a theater. Despite its significance, the Maison du Peuple was demolished in 1965, a loss deeply felt in the architectural community and beyond.

An eclectic example at the crossroads of this new phase in France is the Palais Garnier in Paris, where opulent interiors combine elements of late Baroque, Classicism, and Art Nouveau to create a unique and luxurious space for opera and ballet.

The fin-de-siècle period established Vienna as an influential capital in design and aesthetics through the concept of Gesamtkunstwerk, a comprehensive work of art encompassing various disciplines. The intense cultural and intellectual activity in Vienna during this era, marked by a spirit of innovation and a questioning of traditional values, profoundly shaped the city's artistic and social spheres, with music, opera, and theater playing crucial roles in reflecting societal attitudes and driving aesthetic evolution.

Composers like Gustav Mahler and Richard Strauss pushed the boundaries of musical expression, while the city's theaters and opera houses became arenas for avant-garde experimentation. Playwrights such as Arthur Schnitzler investigated human psychology and sexuality, echoing the investigative spirit of contemporaries like Sigmund Freud, whose exploration of the unconscious mind provided a new framework for understanding the complexities of human behavior and relationships.

Otto Wagner, though not directly contributing to theater typology, played a seminal role in redefining Vienna's arhitecture landscape with his innovative use of materials and assemblies, influencing twentieth-century architecture. His Karlsplatz Station and the interiors of the Postsparkasse exemplify functional scenography, while Joseph Maria Olbrich's Secession Building, completed in 1898, embodies the Secessionist movement's dedication to blending functionality and artistic beauty in contemporary theater design and urban planning.

The avant-garde movements of the twentieth century marked a definitive break from classicism and aimed to lay the foundations for the future modernist phase. Cubism's influence on the performing arts, emerging around 1907 with Picasso and Braque, signaled a pivotal shift towards abstraction and experimentation, impacting everything from stage and costume design to narrative and choreography. The fragmentation of objects and multiplicity of viewpoints inherent in Cubism offered new possibilities for stage design, moving away from traditional scenic realism to abstract, geometric sets that challenged audiences' perceptions and expectations.

For instance, the Ballets Russes, under impresario Sergei Diaghilev from its inception in 1909 until his death in 1929, collaborated with Picasso on set and costume designs. Picasso's work for the ballet *Parade* in 1917 is a notable example, where the Cubist aesthetic was manifested in the costumes and sets, contributing to a groundbreaking fusion of visual art and dance that blurred the boundaries between different artistic disciplines.

Similarly, Russian artist and designer Pavel Filonov, active during the 1910s and 1920s, applied Cubist principles to his theater work, creating non-representational, dynamic sets that reflected the emotional and thematic complexities of the plays. Costumes from this early twentieth-century period, based on Cubist principles, often featured bold geometric patterns, fragmented surfaces, and unconventional materials. These designs transformed the performers' bodies into moving pieces of Cubist art and worked in concert with the abstract sets to create a cohesive visual experience that emphasized underlying themes and emotions rather than aiming for realistic depictions. Additionally, Cubism influenced the narrative structures and choreography of performances, encouraging a departure from linear storytelling by emphasizing simultaneity and the juxtaposition of different perspectives. Choreographers experimented with movements that echoed the Cubist aesthetic, emphasizing the fragmentation of the body and the spatial relationships between dancers, resulting in more abstract and conceptually driven performances where movement, space, and visual design were intertwined to create a multi-dimensional experience.

In the 1910s, avant-garde movements like Expressionism emerged, seeking to convey emotional experiences rather than physical reality. In theater design, this translated into spaces that aimed to evoke intense emotional reactions from the audience, with bold shapes, dramatic lighting, and innovative use of materials. The Großes Schauspielhaus in Berlin, redesigned by Hans Poelzig in 1919, is an iconic example, featuring a cavernous, domed interior and thousands of stalactite-like ornaments that created an immersive, otherworldly atmosphere for spectators. Expressionism's influence extended beyond theatrical spaces to cinema with groundbreaking works like "The Cabinet of Dr. Caligari" (1920), "The Golem" (1920), and "Dr. Mabuse" (1922). These films and their environments introduced a new perception of performance spaces as representations of psychological states, adding a new dimension compared to the grandeur of realistic spaces seen in D.W. Griffith's "Intolerance" (1916).

In Russia, the Moscow Art Theatre's studio adopted Constructivist principles with its use of industrial materials and utilitarian space design, while in the Netherlands, De Stijl's emphasis on geometric abstraction and primary colors subtly influenced set and costume design in performance spaces.

Founded by Walter Gropius in 1919, the Bauhaus impacted all aspects of design, including performance spaces, and became the foundation of the enduring modernism movement that influenced the entire century. Its philosophy of "total" art, combining craftsmanship with fine arts, integrated architecture, furniture, and costume design in theater. The stage workshop led by Oskar Schlemmer explored the relationship between the human body and space, resulting in innovative productions that redefined the performance experience.

Architectural Innovations in Performative Spaces

Since the end of the Second World War, the landscape of performative arts spaces has undergone significant transformations, reflecting changes in architectural philosophies, technological advancements, and shifts in cultural and social dynamics. Several venues have emerged during this period, becoming iconic for their architectural innovation, historical importance, and contributions to the arts. Today, theater architecture embodies flexibility and technological advancement, reflecting the diverse needs of modern performative arts. Contemporary theaters often feature adaptable staging areas, immersive sound and lighting technologies, and designs that dissolve the boundary between performer and audience. These spaces are designed to accommodate a wide range of performances, from traditional theater to experimental performance art, ensuring that theater remains a dynamic and evolving form of artistic expression.

Though not exclusively a performative arts venue, the Centre Pompidou in Paris, designed by Renzo Piano and Richard Rogers, is notable for its radical architectural approach and multidisciplinary focus on culture. Its structural exoskeleton houses a public information library, a modern art museum, and areas dedicated to music and acoustic research, with flexible spaces that have hosted various performance art events, making it a vital center for contemporary artistic expression rooted in exhibited technological innovation.

Frank Gehry's Walt Disney Concert Hall, home of the Los Angeles Philharmonic, was completed in 2003. Praised for its wood-paneled interior and exceptional acoustics, its sculptural juxtaposition of concave shapes and sharp angles has made it a landmark in downtown Los Angeles.

The Sage Gateshead in England, designed by Norman Foster and completed in 2004, features an organic glass and steel structure that emphasizes transparency and openness, inviting the community to explore a microcosm of performative spaces within a single envelope.

References

Adam, J. (Ed.). (1907). *The republic of Plato* (Vol. 2). Cambridge University Press.
Adorno, T. W., & Horkheimer, M. (2002). *Dialectic of enlightenment*. Stanford University Press.
Aristotle, A. (1933). *Metaphysics* (Vol. 1, p. 9). Harvard University Press.
Bachelard, G. (1964). *The poetics of space*. Beacon Press.
Gadamer, H.-G. (2004). *Truth and method*. Bloomsbury Academic.
Halliwell, S. (1998). *Aristotle's poetics*. University of Chicago Press.
Williams, D. (1993). *African dance: An artistic, historical, and philosophical inquiry*. Africa World Press.

Part II
Architectural Space from Embodiment to Movement

The second part of the book is dedicated to an exemplary perspective on performative arts, focusing on ballet, dance, and theater, based on the experience of choreographer and scholar Marco Pelle. Pelle's expertise is observed through three chapters that delve into the intricate relationship between dance and space, the evolution of dance from classical ballet to modern dance, and a conversation with renowned lighting designer Brandon Stirling Baker.

In Chap. 5: "Ballet: Spatial Narrative and Encoded Movement within a Coded Space", Marco Pelle explores how movement bridges the non-physical and physical realms, drawing on Descartes' philosophies. Ballet, as an art form, transforms the stage into a dynamic narrative space where every gesture and position are crafted to convey deep emotional and psychological states. Pelle discusses the evolution of theater architecture from ancient Greek amphitheaters to modern performance spaces, highlighting how these changes have been driven by societal shifts and the innate human desire for transcendence through art. He underscores the role of lighting as an architectural element, with pioneers like Adolphe Appia revolutionizing stagecraft by introducing three-dimensionality through the interplay of light and shadow. By examining iconic scenes such as MacMillan's Romeo and Juliet and Petipa's Bayadere, Pelle illustrates how spatial dynamics and choreography create profound, immersive experiences that resonate with audiences on multiple levels.

The book then traces the evolution of dance from classical ballet to modern dance in Chap. 6: "The Evolution of Dance Through Singular Perspectives", focusing on how shifting societal values and technological advancements have influenced choreographic practices. Pelle emphasizes classical ballet's reliance on a singular, central perspective to create a cohesive visual and emotional impact. Ballet's structured movements and techniques, rooted in the courts of Europe, project ideals of elegance, control, and symmetry. The emergence of modern dance marked a radical departure from these conventions, driven by a desire to express authentic human emotions and experiences. Pioneers like Isadora Duncan rejected the rigidity of classical forms, embracing natural movement and personal expression. Pelle also explores the transformative impact of lighting on dance, with Wagner's innovations in darkness and Appia's architectural approach to lighting redefining the audience's engagement with performance. Through these developments, dance has continually adapted to reflect and challenge the cultural zeitgeist.

In Chap. 7: "A Conversation with Lighting Designer Brandon Stirling Baker", Pelle engages with Baker to explore the symbiotic relationship between light, space, and performance. Baker shares his philosophy that a performance begins the moment an audience enters a theater, with the history and architecture of the space informing the lighting design. He discusses the importance of both light and darkness in shaping the audience's emotional and visual experience, emphasizing that lighting design is about creating a balanced composition of illumination and shadow. Baker's insights reveal how different performance spaces, from traditional proscenium theaters to unconventional venues like the Guggenheim Museum and the Cathedral of Saint John the Divine, require unique lighting approaches. He highlights the adaptive nature of lighting design, noting that each space teaches him how light behaves, allowing him to tailor his designs to enhance the narrative and emotional impact of the performance.

In Chap. 8: "Theater's Infinite Potential: From Grand Stages to Intimate Spaces", Marco Pelle explores the theater as an energetic meeting place filled by artists and the audience. He examines how synergies between artists, the emotions of the audience, music, words, and movements converge within the space. Pelle discusses the 2001 production of Aida at the Giuseppe Verdi Theater in Busseto, directed by Franco Zeffirelli, showcasing how Zeffirelli transformed a miniature stage into an infinite space through clever directorial decisions. By shifting the spectator's point of view, Zeffirelli made the space for the scene seem immense, despite its small size. Pelle also shares his vision of creating a ballet dedicated to the story of Narcissus, where elements of the myth, such as water and air, are part of the audience's journey. He suggests that Venice, with its unique blend of air and water, would be the perfect location for such a performance. The lighting for Narcissus should be reflective, emphasizing the myth's theme of reflection. Venice's luminosity, resulting from light and its reflection on the water, perfectly conveys the myth of Narcissus, representing an environment that is both transparent and defined, much like the story itself.

Chapter 5
Ballet: Spatial Narrative and Encoded Movement Within a Coded Space

Alessandro Melis and Marco Pelle

Abstract This chapter, "Ballet: Spatial Narrative and Encoded Movement within a Coded Space," investigates the relationship between movement, space, and meaning in ballet. Drawing on Descartes' concept of *res cogitans* and *res extensa*, the chapter positions dance as a medium that bridges the internal, intangible mind and the external, material world. Ballet, as a performance art, uniquely utilizes space to manifest inner emotions and abstract narratives, with choreographed movement serving as a spatial dialogue between the dancer's body and the audience. Exploring concepts of spatial relationships and "emotional space compromise," the chapter uses MacMillan's *Romeo and Juliet* as an example to illustrate how spatial dynamics intensify emotional storytelling in dance. Here, Juliet's agonizing distance from Romeo is rendered through deliberate positioning across the stage, amplifying her separation anxiety through spatial tension. This notion extends historically, tracing how performance spaces have evolved architecturally from Greek amphitheaters to Italianate theaters. Each shift reflects both societal change and an increasing desire for theatrical "magic," from the structured light of ancient amphitheaters to the immersive darkness of modern theaters. The chapter also highlights influential figures in theater design, such as Adolphe Appia, who introduced lighting as a dramatic and architectural element. From open-air theaters to enclosed spaces illuminated by artificial light, Appia's innovations emphasized shadows and three-dimensionality, enabling a deeper spatial interplay. Ballet's spatial evolution, from the symmetry of classical spaces to the randomness of Merce Cunningham's non-hierarchical choreography, reflects a shift towards decentralization and audience engagement. In contemporary settings, the chapter notes that digital platforms such as social media embody an experimental performative space where boundaries between performer and audience blur. However, it posits that a physical, utopian space allowing for mutual interaction and a collective gaze remains an unrealized ideal, hinting at future spatial possibilities in the performative arts.

A. Melis (✉)
School of Architecture and Design, New York Institute of Technology, New York, USA
e-mail: amelis@nyit.edu

M. Pelle
Department of Social and Political Sciences, Bocconi University, Milano, Italy

A. Melis and M. Pelle (eds.), *Design for Performative Arts Spaces*,
https://doi.org/10.1007/978-3-031-98215-6_5

Keywords Ballet · Spatial narrative · Encoded movement · Movement and space · Meaning in dance · Descartes · Res cogitans · Res extensa · Inner emotions · Abstract narratives · Choreographed movement · Spatial dialogue · Audience engagement · Spatial relationships · Emotional space compromise · MacMillan's Romeo and Juliet · Spatial dynamics · Emotional storytelling · Performance spaces · Greek amphitheaters · Italianate theaters · Theatrical magic · Lighting design · Adolphe Appia · Dramatic elements · Architectural evolution · Open-air theaters · Artificial light · Shadows and three-dimensionality · Non-hierarchical choreography · Decentralization · Digital platforms · Social media · Experimental performative space · Collective gaze · Utopian space · Future spatial possibilities · Physical interaction · Spatial interplay · Audience-performer boundaries

According to Descartes there are two substances that create space: res cogitans and res extensa.

While Res Cogitans is the non-physical substance of which minds are composed, Rex Extensa is the corporeal, physical substance (in his theory, Descartes states that the latter is created by God) we're all made of, surrounded by and submerged in.

While the Res Cogitans represent the inner landscape, the Rex Extensa represents the entire material world.

We'll later come back to this very two worlds studied by Descartes, because the movement represent the link between those two. And dance, coming from an inner mote, is the most uncompromising act of vassalage to the Rest Extensa.

Space is the result of a compromise. The universal space is the compromise of mechanics, magnetism, forces and much more.

As an artist myself, I will say that the space we live in, create on and modify artistically is the compromise between four very important points: distance, performer(s), movement and the very perception of the three.

In dance we talk about spatial relationship when we create movement that is aware of the frame in which it takes place.

This movement will change its own meaning according to its compromise with the very space.

Let me give you an example: if I want to choreograph a very strong desire between two lovers who are about to be separate, I may decide to have them holding to each other tightly, so that they will look like one only being in the middle of a social frame. I could also separate them and have them running towards each other.

And while we're on the subject of representing separation, let's analyze briefly MacMillan's Romeo and Juliet bedroom scene.

We are in the opening of Act III, inside Juliet's bedroom. The secret newlyweds are sleeping, until the moment Romeo realizes dawn is breaking and he must leave. Juliet, in extreme distress, tries to convince him to stay longer, eve though she's fully aware that he needs to run fast and go into exile.

The order in what the action happens is the following: we see Romeo wearing his cape, walking from the bed (center stage) towards upstage left, where there's the window that he will use to flee.

Juliet awakes and sees him standing on the upstage left corner. She is on the bed at this point, center stage.

Here is what emotional space compromise is: to show Juliet's frantic separation anxiety and pain, Sir MacMillan has her run downstage right, on the opposite and diagonal corner from where Romeo is standing.

And it's from that diagonal distance, the longest geometrical distance that could cover on stage, that she takes two slow steps towards him, before running to him and grabbing his cape to cover herself.

The scene is breathtaking and describes perfectly how she is feeling.

If Sir MacMillan had chosen to choreograph Juliet walking from the bed directly to Romeo (for a total of probably 3–4 steps), the emotional impact and the insightful representation of her feelings would have been completely lost.

Architecture designated to performing space has changed along the centuries.

It's my belief that this change has happened not only according to our society evolution, but also according to the growing need for "magic" that grew withing the performing arts supporters.

We can find the origin of theater architecture in the Ancient Greece's amphitheaters, followed by the Ancient Roman amphitheaters. These can be dated to the VI century BCE. Their design was both evolutionary and revolutionary, if we think that the two oldest existing theatre buildings in India, for example, seem to have been based on Greek models.

Greek and Roman amphitheaters have the iconic half-circle structure with the orchestra space in front of the stage.

Space wise, Greek theaters used to be built on the hillside near sanctuaries, while Romans did not have a specific location for theaters. You can find them everywhere.

Obviously, these funnel-shaped constructions held such an iconic status that will influence for centuries to come the design of buildings of this type.

Little remains of the initial architectural prototypes, but very important details have been reconstructed from the Theater of Dionysus, in Athens.

While the theater was remodeled several times, we know that it reached a capacity of up to 25,000 in the Fourth Century BCE and that it was used steadily up the Roman Period.

The center of the theater was devoted to Dionysus: it was a circular space, containing the altar of Dionysus and used for sacred dances. That space was called "orchestra".

And Dionysus was the reason why theaters where build near sanctuaries: this way, during festival days, the sacred image of the god could be carried into the amphitheater and be "present" at the plays.

In the beginning theatrical representations where not completely free of the religious elements, which means that the attendance to the performance was probably mandatory.

This most likely became the number one worry for the architects of the time: creating a space that could welcome large audiences. And large audiences were the reason why we can assume with a fair amount of confidence that performers performed on a raised platform in order to reach the largest amount of visibility among the audience, while the chorus remained on the orchestra level.

In the history of architectural evolution, we should mention Teatro Olimpico in Vicenza, Italy, designed by the great Andrea Palladio and built between 1580 and 1585.

Teatro Olimpico is the first permanent indoor theatre of the Italian Renaissance and it's clearly inspired by the ancient Roman architecture.

We must mention in this section the last evolutionary step in teather architecture: the Italian-style theater. It's the type of theatrical structure characteristic of Italian architecture that was particularly widespread between the seventeenth, eighteenth and nineteenth centuries.

The architectural peculiarities are to be found mainly in the following features:

(1) The hall (now called "orchestra") went from being rectangular, typical of court theater, to reach a horseshoe shape, creating an orchestra seat area that will become over time, from a space intended for dancing, a place designated for a better visibility of the performance.
(2) The elimination of the tiers of seats ("steps") in favor of the construction of boxes separated from each other and divided in height by orders;
(3) A greater depth of the stage to allow the use of the innovative perspective wings and the possibility for the actor to act inside and not in front of the stage, as was customary in Renaissance theater.

If we look at the most evolved Greek and Roman theaters, we can say that they feature most of those architectural elements except a few, one more important than all others. There's a stage, an orchestra, seats that allow visibility and capacity. They even had tents build on the "skené" (scene), where actors would change. They only lack one architectural element: the light.

What do I mean by that? That the ancient Greeks and Romans build their theaters strategically so that they could harness the full potential of the sun. Usually, they were built in such a way that the stages would face the sun that rose behind the audience.

This design allowed the sun to hit the stage and the audience to enjoy the show. Different times of the day would allow for different kind of natural lighting.

I was a guest of the popular Lake Tahoe Dance Collective festival, which, like the ancient Greeks and Roman, only uses the natural light. My request for one of the pieces I created for the festival, was for it to be performed right before the twilight, to give that pas de deux the "nocturne" quality it required.

One small problem with using sunlight: not all audience members can deal with the strong and vibrating light of the afternoon Californian sun and, you cannot rely on the sunlight, due to possible disruptive weather conditions.

One last thing: when I mentioned lighting before, I talked about it as an "architectural element".

It is. In order to make a bidimensional element three-dimensional, we need objects on stage along with light and shadows.

It's interesting that the man who is considered a pioneer in theater lighting was an architect, precisely a swiss architect by the name of Adolphe Appia. Most famous for his work on the operas by Richard Wagner, who was the first one to introduce the audience to complete darkness in his newly built theatre in Bayreuth on 13 August 1876, during the Bayreuth premiere of Das Rheingold and the world premiere of the complete Der Ring des Nibelungen.

Adolphe Appia rejected bidimensional elements of stage (painted backdrops i.e.) and considered shadows as important as light, to create a connection between the actor, the stage and the space.

How did we go from sunlight to Adolphe Appia?

Well, when theaters became enclosed, it was necessary create buildings that had windows, so that some light could light the stage and Italians introduced the use of candles on and off stage.

Candles were introduced in Italy in the late 1500's by another architect: Sebastiano Serlio. He created sort of flasks filled with colored water, placed in front of candles. This would create a colorful environment but, again, not manageable intentionally. It also required a very large number of chandeliers to deal with all the candles placed on the edge of the stage.

Another architect was essential for theater lighting: Nicola Sabbatini (1574–25 December 1654). Mr. Sabbatini brought dimming into the stage, and created a mechanism to direct a spotlight, to bring light to specific parts of the stage. He was probably the first man who saw light as a dramatic element that could work along with the play or with the opera.

Adolphe Appia brought the last evolutionary step into the architectural element that light and shadows are: three-dimension. He understood that now that light was manageable and directable, so had to be shadow.

With him we enter the modern era of theater lighting. More about him and his revolutionary vision on Chap. 7.

I personally consider Baroque the era when the concept of magic starts to enter the theater. From the development of the music to the development of the vocal technique (which will see its apex in the Bel Canto era), to the furthering of the ballet technique, everything screams MAGIC!

It becomes clear that performances aim at a collective trance, that they become transcendental experiences. As the voices go higher so do the dancers, who start to try to elevate themselves beyond their feet.

During the French Revolution, heels were forbidden; therefore, many choreographers looked for solutions to add length, lightness and etherealness to dancing.

Charles Didelot, in Paris, invented a "flying machine" that was able to lift a ballerina upward and pose her on her toes (pointe) before leaving the ground again.

From the flying machine to pointe shoes, it was a short step (no pun intended). Within a few years, ballerinas would dance on pointe, adding magic, lightness and height to each of their steps.

In 1877, just one year after the opening of the Bayreuth Festival, with darkness being introduced to the audience for the first time, a ballet titled "Bayadere" made its debut at the Imperial Bolshoi Kamenny Theatre in St. Petersburg, Russia. The ballet was choreographed by French choreographer Marius Petipa on music by Ludwig Minkus.

The reason why I write about "Bayadere" is its second act, known as *The Kingdom of the Shades*. During the second act, Solor, who's in love with beautiful bayadere who has just recently died, takes solace in dream induced by opium. During this dream he finds Bayadere in the underworld where the shades of dead women reside.

And here's the genius idea of Petipa: he was able to give Russia its own iconic style, a mixture of French steps into the grandeur of the Russian idiom, from its architecture to its own icons.

Let's see what happens during Solor's dream: she appears to him as a spirit from the heights of the Himalayas, called The Kingdom of Shades, accompanied by the ghosts of her fellow temple dancers (a total of 32). The staging we see was inspired by Gustave Doré's illustrations of Dante's *Paradiso* from *The Divine Comedy.*

Homans in "Apollo's Angels—A History of Ballet" (2013) writes: "as a single shade, in white tulle and draped in gauzy veils, steps onto an empty and brightly lit stage from the far upper-right corner. In profile, she takes an elegant, forward-reaching arabesque with her leg lifted high behind her, followed by a deep reclining back bend and two steps forward. She repeats the sequence again as another shade emerges from the wings and follows in synchrony, then another, and another. One by one, as if to infinity, along chain of shades (sixty-four, to be precise, later reduced to thirty-two) wind their way single file across the stage and back, tracing a serpentine path and advancing steadily to make room for the next row. The visual crescendo builds with each repetition until the stage is full and serried ranks of dancers pose in perfect formation."

What we witness, when we look at the Bayadere's Kingdom of Shades is the most iconic scene in ballet, the one that made ballet what it is today: grand, magical, inspiring.

We can see the powerful and grandiose Russian courts, with their tsars. We can imagine them walking in a procession through the endless ballrooms of their palaces.

At this point we are in front of the most intelligent and meaningful use of space in ballet: it's no longer space used to perform, but space used to define a way of thinking, a way of ruling.

It's a space turned into a Country, into a culture, into history.

Bayadere is the first ballet of its kind to do that.

Couple of decades after Bayadere, choreographers began take serious consideration of the space that the very dancers represented. How?

For example, the great Agrippina Vaganova, a Russian ballet teacher who developed the Vaganova method—a technique developed from the teaching methods of the old Imperial Ballet School, invented what's known today as the "Vaganova Square".

The Vaganova Square is an imaginary layout around the dancer that has 8 points. Odd numbers are the mirror (number 1) and the three walls around the dancers (numbers 3–5 and 7). The corners of the room are the even numbers.

The audience is *always* number 1. So, no matter the point the dancer is facing while dancing, they need to be aware of their number 1, because that's where the audience's point of view is.

And hierarchy is part of performing: there's a hierarchy in the roles that are played and danced, but there's also a hierarchy of space and its use. So, Vaganova's point number 1 will be the most important direction to face, because that is where the audience sits.

The choreographer that will break this hierarchy, going against all codes and rules will be Merce Cunningham. Mr. Cunningham, a former Martha Graham's dancer, broke all rules when it came to symmetry and vision priority. Dancers face every direction and that very direction is their number 1 direction, it's their audience.

This asymmetrical and non-hierarchical method adds up to props that give a sense of randomness, of wilderness almost, that became Mr. Cunningham's signature.

John Cage's music added to the math, creating something that was between an experiment and an incredible masterpiece of craftmanship.

One of Merce Cunningham's most iconic works is Rainforest, Andy Warhol agreed to let Cunningham use his installation titled "Silver Clouds", made of several Mylar pillows filled with helium, so that they floated freely in the air.

The element of lack of controllability and the random relations that Mr. Warhol's floating pillow bring on stage is mesmerizing.

Because there are no directions and no hierarchical element, the feeling the audience must be completely immersed in the scene in a sort of a surrounding effect.

Space has evolved over the centuries according to logics that have gradually shifted the public's gaze, from a dominant position in relation to artistic creation, to a slightly more independent position.

Of course, the funnel-shaped theatrical form remains one of the most democratic spaces when it comes to the domination of the gaze. However, it does not work well for classical dance, which was created between the 1500 s and 1800s based on optical illusions that only work in a horseshoe-shaped theater.

The Italian-style theater, with the stalls and the boxes, reflects, in a theatrical form, the hierarchical structure of society: the wealthier gaze has the best view. The classes that can afford less, financially speaking, are relegated to the higher boxes, where the sound is good but the view is completely disproportionate and unsuitable for certain types of performances (especially dance).

All other spaces represent cultural and artistic niches, attempts to bring art outside the walls built specifically for them. But they stop there. They do not go beyond.

Are there spaces worth discussing? Probably not, as they are all minor.

However, it is worth understanding that in the Western world, the construction of performative space has followed a logic parallel to that of social and especially religious structures.

In primitive times, everyone would gather around the fire, a vital and sacred element. The bonfire symbolized salvation from darkness, from animal attacks, the necessary element to cook game, for warmth, for seeing in the dark, and at the same time, the sacred light to turn to in moments of despair.

Thus, the first theaters, and this continued for hundreds of years, had a similar circular shape, where instead of the fire, we now had the artists, with their sacred or popular art. The gaze of the public was, in a sense, circular.

As humanity evolved, so did the concept of the sacred, and the concept of participation in the sacred and in society. The space became rectangular, divided into three segments: that of the faithful, that of the celebrant, and that of God. Seen in a secular light, that of the people, that of the leader, and that of the administrators.

Thus, the theater evolved: it became a rectangular space of three segments: that of the audience, that of the music, and that of the actors. Or that of the audience, that of the actors, and that of the mystery (behind the scenes).

This parallel should not be underestimated, because to this day we have not yet imagined a new system of praying, other than through a leader of our religious creed, or of governing, other than through a leader of our political creed, or of observing, other than through an icon that takes shape and performs before our eyes.

The gaze remains rectangular, in a sense.

The only evolution that has occurred is that of occasionally shifting the direction towards which the artist performs. But not their position.

It's either central or on an altar, in a sense.

We will know we have taken a step forward the day our gaze is elevated, parallel and central at the same time, alongside that of the artist.

That will be the moment when the audience will be both stage and seats, altar and sanctuary, elected and elector.

Now, the only place like what has been just described above is social media: that's the place where the entertained is its own entertainer, the devotee its own icon, and the voter its own elected.

But it's clear that we are still missing a physical place for what is currently pure utopia and that it will probably require many more years before becoming a reality and an harmonious, fully balanced place.

Reference

Homans, J. (2013). *Apollo's angels: A history of ballet*. Granta Books.

Chapter 6
The Evolution of Dance Through Singular Perspectives

Alessandro Melis and Marco Pelle

Abstract "The Evolution of Dance through Singular Perspectives" explores how classical dance evolved from an aristocratic art form into an expression of personal and emotional movement, tracing the influence of spatial perspective and lighting on the art of dance. Ballet, originally centered around the spectator's fixed viewpoint, embodies elegance and structured beauty through carefully orchestrated movements, such as en-dehors and crossed forces (*forces croisées*), which convey symmetry and poise. This singular, central perspective optimizes the visual effects of classical dance, reinforcing its courtly origins that celebrated political and social hierarchy. However, as the world entered the late nineteenth century, modern dance emerged as a rebellion against ballet's rigid beauty and the idealized portrayal of fantastical realms. Influenced by the Industrial Revolution and shifting societal roles, modern dance began to challenge not only classical form but also the theater's spatial limitations. Pioneers like Isadora Duncan advocated for the authenticity of personal movement, dancing barefoot and in natural settings, expressing a liberated self-image that rejected conventional constraints.

Keywords Evolution of dance · Singular perspectives · Classical dance · Aristocratic art form · Personal movement · Emotional expression · Spatial perspective · Lighting influence · Ballet · Fixed viewpoint · Elegance · Structured beauty · En-dehors · Crossed forces · Symmetry · Poise · Courtly origins · Political hierarchy · Social hierarchy · Modern dance · Late nineteenth century · Rebellion against ballet · Rigid beauty · Idealized portrayal · Industrial revolution · Shifting societal roles · Theater limitations · Isadora Duncan · Authenticity of movement · Barefoot dancing · Natural settings · Liberated self-image · Conventional constraints · Dance as expression · Movement innovation · Dance history · Artistic evolution · Cultural shifts · Performative arts · Dance technique

---

A. Melis (✉)
School of Architecture and Design, New York Institute of Technology, New York, USA
e-mail: amelis@nyit.edu

M. Pelle
Department of Social and Political Sciences, Bocconi University, Milano, Italy

A. Melis and M. Pelle (eds.), *Design for Performative Arts Spaces*,
https://doi.org/10.1007/978-3-031-98215-6_6

Classical Dance requires the spectator's point of view to be singular: central to the stage, positioned slightly below the dancers. This perspective is optimal to fully appreciate the visual effect created by classical dance.

In fact, ballet originated as an art form within courts, showcasing the political significance of various court figures, evolving into an art form designed to amaze with plastic poses, high legs, pointe work, and technical virtuosity such as jumps and spins.

All these virtuoso movements are based on a series of crossed forces (appropriately called "forces croisées"), which create the illusion of length, symmetry, and the relationship between different parts of the body engaged in movement.

Through these crossed forces, the use of en-dehors (turnout in English), and other techniques, the dancer's body becomes an image of elegance, self-mastery, romantic yearning, and emotional momentum, distancing itself from the type of performance and circus-like showmanship.

However, to achieve all this, a theater that acts almost as a screen for the spectator is needed: depth is given by shadows and props, but anything lateral and peripheral is effectively avoided. Movement is never viewed in profile, because it would be like admiring an embroidered dress displayed inside out.

It is only with the birth of modern dance that we begin to consider spaces different from those defined by classical arts.

Modern dance emerged as an artistic rejection of classical dance, a form of rebellion against the image of a world that, by the end of the nineteenth century, no longer resonated with the new generations.

The Industrial Revolution, along with the decline of Victorian moral and aesthetic structures, paved the way for what became a true artistic and cultural revolution. Women no longer saw themselves in the narrow image of the faint, magical angel of the hearth. The era of conventional, vapid beauties was over. Dance had to return to representing human emotion, not just the fanciful imaginings of woods and magical creatures.

Additionally, more and more dancers felt the need to express their own movement, to create it specifically for and by themselves.

Until almost the entire nineteenth century, the choreographer was a sort of puppeteer, shaping dancers according to their own taste and artistic vision.

With industrialization and the desire for new discoveries, young dancers found it impossible to submit to the singular will of the creative arm. They rebelled.

The first to start a true rebellion (albeit by looking backward rather than forward) was Isadora Duncan, the Californian artist born in 1877 and unanimously considered the mother of Modern Dance.

Why such an important title? Isadora Duncan rebelled against everything: the affected classical elegance, dancing on pointe, wearing corsets and tutus. Clad only in a white tunic, with bare feet, Isadora Duncan brought us back to the naturalness of movement, asserting her right to create a dance made in her own image.

Among the various ways Duncan's rebellion manifested was dancing in contact with nature, far from the sacred places of choreographic art.

For instance, she danced in the gardens of Chesterwood, in Stockbridge, Massachusetts.

Owned by sculptor Daniel Chester French (1850–1931) and his wife, the performance was paid and attracted all the high society figures willing to pay $2 to see the Botticellian beauty dance among the trees of the property.

Another significant step in the evolution of theater, and consequently dance, was the use of lights. Or more precisely: the management of darkness.

The architectural space, until then used passively, became, towards the end of the nineteenth century, a place that managed not only the characters on stage but also the emotions in the audience.

For while it is true that the dancer amazes the eye, it is equally true that light amazes the soul. And darkness moves it.

Until over the mid-1800s, theaters were illuminated by candles. Over time, candles gave way to oil lamps, used with chandeliers and standalone units, mostly placed in glass containers.

Turpentine oil, petroleum, and other combustible fats were used as lighting materials.

We must, however, wait for Wagner to understand the fundamental importance of an aggregating theatrical element: darkness.

In total darkness, the spectator is immersed in a mysterious and parallel universe: that of the performance. Only darkness can bridge us and the stage. Wagner was the first to grasp its fundamental importance.

In his writings, Wagner often lamented the bidimensionality of the direction of his works: there was the painted backdrop, with an obsessive quest for realism and grandeur, and his music. Wagner, with his genius, understood that the flight of his music was not supported by scenic flight. However, he did little to challenge the status quo of a simplicity in stagecraft that he so strongly disliked.

His only real revolution, at the scenic level, was to construct an extremely low and hidden orchestra pit (mystic gulf), so that the Bayreuth theater was almost completely dark during performances. It was the first time the audience found itself almost completely immersed in darkness.

But that was all.

The one who truly revolutionized the concept of space through lights was a Swiss architect named Adolphe Appia (1862–1928).

And if Wagner understood the importance of darkness in theater, Appia understood the importance of shadows on stage, and his work, revolutionary and surprising, approaches more the work of a sculptor than that of a lighting designer.

Appia realized that the presence of three-dimensional objects on stage created shadows and hues of different shades, and managing these shadows also meant managing the emotional journey of the audience. Nothing, from Appia onwards, would ever be the same. Once three-dimensionality became part of the stage, the audience became a fundamental part of the theatrical work, no longer a passive element. The gaze changed, and the visceral nature of the scene acquired a new level, a new depth.

Chapter 7
A Conversation with Lighting Designer Brandon Stirling Baker

Marco Pelle

Abstract In this chapter, Marco Pelle interviews lighting designer Brandon Stirling Baker, whose work spans prestigious ballet companies, Broadway stages, and iconic museums. Through their conversation, Baker explores the nuanced relationship between architectural space, light, and darkness, emphasizing how each space "breathes" and informs his design approach. Baker reflects on the power of storytelling, asserting that a performance begins with the audience's first encounter with a space's unique history and design. His approach to lighting varies with each setting, from creating intimacy in Broadway's St. James Theater to achieving epic scale at the Park Avenue Armory. His work also extends to unconventional venues like the Guggenheim Museum's Rotunda and the Cathedral of Saint John the Divine, where the crafts ambiance uses reflected light and shadowless illumination. Baker's designs pay homage to each space's character, merging architectural form with the emotional journey of performance.

Keywords Interview with Brandon Stirling Baker · Lighting design · Architectural space · Light and darkness · Design approach · Storytelling in performance · Audience experience · Unique space history · Broadway lighting · St. James Theater · Epic scale · Park Avenue Armory · Unconventional venues · Guggenheim museum · Rotunda lighting · Cathedral of Saint John the Divine · Ambiance creation · Reflected light · Shadowless illumination · Space character · Merging form and emotion · Performance journey · Nuanced relationships · Design variability · Crafting atmosphere · Emotional impact · Visual storytelling · Theater design · Museum spaces · Impact of light on performance

I met Brandon Stirling Baker, an internationally renowned lighting designer whose lights illuminate companies including the New York City Ballet, the American Ballet Theatre, the Boston Ballet, Alvin Ailey, La Scala, the Paris Opera, and many others. He was also nominated for the Tony Awards in 2024 for the musical "Illinoise". I wanted to talk to him about architectural space and light, so important for theater.

M. Pelle
Department of Social and Political Sciences, Bocconi University, Milano, Italy

A. Melis and M. Pelle (eds.), *Design for Performative Arts Spaces*,
https://doi.org/10.1007/978-3-031-98215-6_7

Q—So let me ask you, first of all, what is a performing space for you? How do you bring it to life?

A—In my opinion, the moment an audience walks into a theater, the statement begins. And what I mean by that is that the story begins the moment we put one person on a stage. It becomes a story. There's a very famous quote from George Balanchine. He said "Put a man and a girl on stage and there is already a story; a man and two girls, there's already a plot." In any space we are in there's already a story going on. There's history in that room. When I just opened the show called *Illinoise* on Broadway, the very first day of our technical rehearsals, the general manager of the St. James Theater came over and explained to us the history of the space, the history of the theater and how the history is always with us whether we like it or not and how that history inspires us even if it's subconscious or if it's conscious or if we're aware of it.

But there's something beautiful about the history of what has happened in a particular space.

Q—Does that have an effect on how you will design the lights?

A—If I'm being very specific on how I want to treat a space, as a lighting designer, I am immediately inspired by the height of a space and the width of a space. And that teaches me how a space *breathes*. If a space is very tall, as a designer I like to use the height of a room. If it's very wide I like to use the width. This means that the light comes from its highest point so that it feels like it's coming from the heavens or feeling feels like the light is coming from beyond this world.

For example, when I work in theater shows or theater projects where the audience is in the round, in a circular formation, it becomes a very communal experience and so the light has to somehow be universal, something that as a collective as an audience we experience and share together. And that to me is directly, first and foremost, inspired by the space itself.

Q—How important is darkness in your work? And how do you think darkness affects the ability to turn a space or a work into a statement?

A—Well, as a lighting designer, don't just design the light. We design the darkness. We design what you see and how you see it and we control how much of the space you can see. That is equally important as the light (or the lack of light in your case). The darkness is a part of the composition. It is compositional thinking. I think when you when you listen to music, you're demonstrating how you hear, how you listen. As a lighting designer, I demonstrate how I see.

That goes for what is light and for what is dark. How I understand that or how I decide that is to the overall mission or the overall story that's being trying to be told on stage. I'm always aware on how much light exists in this world.

In much of my work lately the light is not symmetrical, is not coming from both stage left and stage right. It's maybe coming from one direction only. The reason for that is I find a lot of beauty in the imperfections of life.

Q—How does that work out with a space? Theatrical spaces tend to be symmetrical, and choreographic works tend to be symmetrical most of the times, I would say.

A—I would argue that there is no normal. The shape of a space inspires the shape of the light for me.

If the dancers are all symmetrical, that's a conscious decision. That's a choice. So there are sometimes that that's it. It's asking for symmetry in the light, and you shouldn't fight that.

That decision, though, comes from my conversation with the director or choreographer. Are we supposed to view this symmetry in abstract way? Are we supposed to view this symmetry in a very clear, crisp way? I ask, "How do we want to view this world?" Once I understand the world, I can understand the light.

There is no copy and paste. Let's take the Broadway show "*Illinoise*" as an example: I've done it now in four different spaces, I've created the light in four different spaces, drastically different.

The very first one was upstate New York at the Bard Fisher Center, a traditional proscenium theater. Audience is on one side; the stage is on the other. It's framed by a proscenium, and it's a very traditional presentation. After that, we took it to Chicago. And in Chicago, we also were in a proscenium theater.

However, the audience was much closer. It felt like they were surrounding the stage. It's almost like a thrust. And thrust being like a horseshoe shape.

And then the third version was at the Park Avenue Armory, which is like a warehouse. That was almost an arena version of our show. And even if you know nothing about the show, you don't have to. What's important to note here is the space behaves differently.

Like, it changes the way you view a performance. At the Armory, the audience is far and they were viewing it from almost above.

It's like as if you're watching the show from Madison Square Garden.

That's the experience. And then when we brought it to Broadway, it was a return to a small space And what I learned from this experience is that light is not created equal in different spaces. So an idea that would read in the Armory as, epic and larger than life may be overkill and too much in a smaller theater.

The way I like to phrase it is that I'm a designer of feeling, and also a designer of context.

On Broadway, it was more intimate and I could create the sense of a real campfire, the intimacy of a campfire. So, there were entire scenes that were only lit by one light, but I could only achieve that in a successful way on broadway whereas in the Armory I needed a lot more light to achieve the same idea.

Q—You also copllaborate with museums. What is your what is your approach to spaces like those, so different from regular theaters?

I have a long partnership with the Guggenheim Museum here in New York, and I've worked with it's there's a program called the works in process series. And for many years, I used to light their Rotunda shows. The Rotunda is the main gallery space, and they would do these one-time only performances in the Rotunda. For this process, you have only thirty minutes to set up the lights. It's extremely fast. And the reason for it is because it's a museum during the day. It's not a performance space.

There's a lot of planning that must go into that. So how are you gonna use these thirty minutes?

A—I have done, over time, three of these projects. I learned that less is more. And I also it goes back to the fact that a room or a space can teach you how light exists. So, I would go to the museum as a just a as a regular viewer, just to see art and see how natural light behaves in the space. And in the Solomon R Guggenheim Museum, the main source of light is a giant skylight at the very top of this theater. Almost all the entire museum itself is lit by natural daylight coming through the skylight, supported by some track lighting and normal lens. But the main source is from above. Skylight. It's natural light. However, my show was at nighttime.

Hence, no natural light. So, what I decided is this: the entire space is a white room. A bleach white, amazing architecture that's iconic, very famous. I therefore decided that the entire room was gonna be lit by reflected bounce light.

I took inspiration from one of my favorite artists, James Turrell, and I placed two lights in this theater, two very big lights, and I lit the entire ceiling with deep blue light, and that alone lit the entire space. Instead of light coming through the skylight, it focused up, and they were shot it reflected light down to the floor.

Q—What was the effect?

A—I learned from the work of James Turrell that reflected light can really be enough. And even the room that we're in right now, in your living room, like, the light that's lighting us is reflected light. I can see you perfectly well.

And it's not directly hitting us, but it's bouncing off the floor. It's bouncing off the window. This is no different from what I did at the Guggenheim. All the light does is lighting the ceiling, and the audience's eyes adjust. Your eyes start to provide clarity. So, for this Guggenheim performance series, truly the entire project was lit by two lights in this massive room. And what was so beautiful about it, it's that it was shadowless light. It was light that had no shadow.

Q—What's the most unconventional space you lit?

A—I would say it's the Cathedral of Saint John the Divine.

I lit an opera there. It was a collaboration with Philip Glass and the opera singer Anthony Roth Costanzo (countertenor). It was a show called *Glass Handle*, with the music of Handel and Philip Glass put together. Lighting that space was very complicated.

The space itself is a character. The space itself has history. But again, like for any space I work in, I will look and see how does light already exist. Not to say, do I just follow that, but I can learn from it.

So at the Cathedral of Saint John, the divine, a lot of the light was for the columns. It architecturally followed the logic of the space. But in addition to that, kind of in a similar spirit to how I looked at Guggenheim, a lot of the light was bouncing off the ceilings and reflected on to the different stages. And I tend to do that, I think it's almost like my love letter to a space. It's about me paying tribute to the room I'm in.

Chapter 8
Theater's Infinite Potential: From Grand Stages to Intimate Spaces

Marco Pelle

Abstract The chapter examines how theater and performance spaces transcend physical limitations, adapting to amplify artistic expression and create profound audience experiences. It begins with Franco Zeffirelli's 2001 production of *Aida* in the tiny, 307-seat Giuseppe Verdi Theater, where Zeffirelli ingeniously reimagined the grand Triumphal March scene, using spatial illusion to evoke grandeur. The author reflects on his own imagined ballet on the myth of Narcissus, proposing a circular, reflective stage on Venice's canals—symbolically evoking Narcissus's self-reflection and the element of water central to the myth. The chapter concludes by exploring Venice as an ideal venue, where light and water's natural interplay embody Narcissus's fateful reflections. Through these examples, the chapter illuminates theater's potential to transform spaces, enabling narratives to connect deeply with audiences, whether through the vastness of myth or the intimacy of personal journey.

Keywords Theater spaces · Performance adaptation · Artistic expression · Audience experience · Franco Zeffirelli · Aida production · Giuseppe Verdi Theater · Triumphal march scene · Spatial illusion · Grandeur · Narcissus myth · Reflective stage · Venice canals · Self-reflection · Water element · Ideal venue · Light and water interplay · Fateful reflections · Narrative transformation · Deep audience connection · Vastness of myth · Intimacy of personal journey · Symbolic evocation · Spatial dynamics · Creative reimagining · Theatrical innovation · Reflective design · Environmental influence · Immersive experience · Mythical narratives

The theater is also an energetic space in the sense that it is not at all empty but rather a meeting place filled by artists and the audience. In this space, the synergies of the artists, the emotions of the audience, the music, the words, and the movements all converge. The theater is a space filled with everyone present.

M. Pelle (✉)
Department of Social and Political Sciences, Bocconi University, Milano, Italy

A. Melis and M. Pelle (eds.), *Design for Performative Arts Spaces*,
https://doi.org/10.1007/978-3-031-98215-6_8

But what happens when the space is so small that it fills up quickly, running the risk of not being able to convey the artistic energies due to its size?

A very interesting case is the 2001 production of Aida at the Giuseppe Verdi Theater in Busseto, directed by the great Franco Zeffirelli.

The theater, located within the Rocca Pallavicino in Busseto, seats 307 people and has a stage width of 7 m.

It is practically a miniature. In 2001, Zeffirelli accepted the challenge of directing Giuseppe Verdi's most opulent opera: Aida. With a cast requiring the presence of a significant chorus and one of the most famous scenes in opera history, the Triumphal March, Aida typically demands very large theaters to stage the splendor of ancient Egypt. Notable productions include those at the Arena di Verona, the Metropolitan Opera in New York (where even horses parade), and the opening in Cairo in 1871, where even elephants paraded.

The second scene of the second act opens with the celebration of Radames' victorious return from war. The "Triumphal March" is a triumph of music, parades, and splendor. There are priests, priestesses, dignitaries, soldiers, slaves, and common folk. The stage fills with extras and choristers supporting the main actors in what is, without a doubt, the most lavish scene ever created in opera history.

How can this grandeur be translated onto a 7 m stage?

Here we see the genius of Maestro Zeffirelli.

During this important parade, Zeffirelli shifts the spectator's point of view from the privileged position of someone watching the parade from the front row to someone who is present at the parade but cannot see anything because they have all the spectators in front of them.

In practice, we witness a diagonal entrance (from downstage right to upstage center) of Radames followed by Amneris.

Once they reach upstage center, they are swallowed by the crowd, which, facing the backdrop, obstructs the audience's view. Zeffirelli makes us imagine that the triumphal march is taking place where the backdrop is, but we, the audience, cannot see anything because we have a dense group of people jumping with joy while cheering for the victorious Radames, who is marching amid the crowd's exaltation.

This directorial decision was brilliant: we do not see the triumphal march but perceive it, because, as often happens at the Halloween Parade in New York, we did not arrive in time, we have a sea of people in front of us, and the only thing we can do is imagine the parade through the music we hear and the glimpses we catch.

In this case, by creating a visual impediment, the director made the space for the scene seem infinite: he knows the truth. The audience perceives an immense space.

The theater is also this: the transformation of space from large to small, but also from small to infinite.

Personally, I have worked on the idea of creating a ballet dedicated to the story of Narcissus, where the elements present in the story are part of the audience's journey.

In the story of Narcissus, two elements strongly recur: the first is water, and the second is air.

Narcissus, the son of the river god Cephissus and the nymph Liriope, was of such beauty that anyone who saw him fell in love with him.

When Narcissus was an infant, the seer Tiresias told Liriope that the boy would live to old age only if he never "knew himself."

When Narcissus was 16, he was seen by Echo, a poor nymph who had been punished by Hera, Zeus's wife, for distracting her with frivolous stories to prevent her from noticing her husband's infidelities. The punishment consisted of taking away almost all her speech. The only thing she could do was repeat incessantly the last words spoken to her.

When Echo saw Narcissus, she fell madly in love with him, followed him into the woods, and after an initial encounter where she could only stupidly repeat Narcissus's questions, she was harshly rejected by him. Heartbroken, Echo was left with nothing but to stupidly repeat the words of rejection, a sorrow that consumed her until she became a cave from which the echo was born.

This is the first element of the myth: air.

The second is water.

Nemesis, according to some the daughter of Zeus, moved by Echo's cruel sorrow, decided to avenge her. She made it so that Narcissus would come across a crystal-clear spring and bend down to drink. Faced with the water's reflection, Narcissus saw himself for the first time and fell in love.

He fell madly in love with himself, at first not realizing that he was the image he saw and then fully understanding the curse he was destined for. This was a true awakening. Let us not forget that Narcissus comes from Nàrkè, which means "sleep" and "numbness." Narcissus's is an authentic awakening. An awakening that would lead to his death.

Once he understood his curse, Narcissus let himself die in despair.

I believe that the perfect space to tell the myth of Narcissus is Venice. Venice is a city where the elements of air and water blend indelibly, creating a mix of unmistakable sounds and reflections.

The Narcissus should be represented within a circular space, where people can follow the story of the boy of fierce beauty and the characters surrounding him. This space should be placed directly on the water, representing both the birth and death of Narcissus, and recreate the ability, typical of Venetian architecture, to return sounds to those who traverse it.

If we were to create a space to represent Narcissus, it should be a circular, glass-covered platform, where the architecture assumes the same circularity of the story, and where the audience returns, at the end of the myth, exactly to the starting point, to the birth of the myth itself.

In fact, Narcissus is born from water; his mother is a freshwater nymph, and he meets his death through water.

It is inconceivable to separate the myth from its primary element.

The lighting for Narcissus should be, as we heard from Brandon Stirling Baker, reflective light. Because it is precisely around the reflection that the entire myth of Narcissus revolves. He sees the reflection of his beauty through the eyes of those who look at him, sees his beauty through the reflection in the water, and Echo knows the words of rejection through the foolish repetition of Narcissus's last words.

Everything in the myth speaks to us of reflection: because it is only through it that we see the objectivity of reality.

Venice is a city whose luminosity results from two elements: light and its reflection on the water.

No place can convey the myth of Narcissus better than the magical city par excellence. The awakening of the myth also represents our awakening within an environment that is both transparent and defined, immersed in the city that has made reflection its most exquisite characteristic.

Part III
Case Studies and Best Practices

The third part of this book explores practice-based research within the realm of performative arts, illustrating the vast array of possibilities that this field encompasses. This part presents an extensive range of perspectives, beginning with reflections on historical case studies and extending to contemporary experiences shared by the book's authors. Additionally, it includes insights from a carefully selected group of guest experts, practitioners, and scholars who contribute their expertise and theoretical activities.

By showcasing both current and future opportunities, this part not only highlights the diversity and richness of performative arts but also lays the groundwork for a comprehensive manual. This manual is designed to be an invaluable resource for professionals, researchers, and students within the field of university education, fostering a deeper understanding and appreciation of the performative arts through both theoretical and practical lenses.

Chapter 9
Theaters of Pisa

Alessandro Melis and Marco Pelle

Abstract This chapter, *Theaters of Pisa: The Enlightenment Influence on Tuscan Theaters* explores how Tuscan performative spaces, particularly in Pisa, became cultural and political beacons during the Enlightenment. The Lorraine family leveraged Pisa's theaters as laboratories for progressive ideals, exemplified by symbolic architectural transformations that mirrored Tuscany's social reforms, including the abolition of the death penalty. In the Regency Years, the city embraced Baroque-inspired ephemeral architecture for royal celebrations, imbuing spaces with grandeur that reflected Enlightenment values. Key projects, such as the Rossi theater, became both social equalizers and symbols of Enlightenment ideals, embodying the transformative power of theater to bridge art and education. By the Risorgimento, Pisa's theaters—especially the grand Teatro Verdi—evolved into civic spaces representing Italy's burgeoning national identity, blending public accessibility with architectural sophistication. This chapter contextualizes Pisa's theaters as evolving symbols of Tuscany's rich cultural heritage and socio-political progress.

Keywords Tuscan theaters · Enlightenment influence · Cultural beacons · Political ideals · Lorraine family · Progressive ideals · Architectural transformations · Social reforms · Abolition of death penalty · Regency years · Baroque architecture · Ephemeral architecture · Royal celebrations · Grandeur · Rossi theater · Social equalizers · Transformative power of theater · Art and education · Risorgimento · Teatro Verdi · Civic spaces · National identity · Public accessibility · Architectural sophistication · Cultural heritage · Socio-political progress · Symbolic representations · Performative spaces · Tuscany's legacy · Enlightenment values

A. Melis (✉)
School of Architecture and Design, New York Institute of Technology, New York, USA
e-mail: amelis@nyit.edu

M. Pelle
Department of Social and Political Sciences, Bocconi University, Milano, Italy

A. Melis and M. Pelle (eds.), *Design for Performative Arts Spaces*,
https://doi.org/10.1007/978-3-031-98215-6_9

The Enlightenment Influence on Tuscan Theaters

The transformation of spaces for the performative arts in Tuscany, and particularly in Pisa, highlights the social and political influence of these architectural types during periods of significant political change. Tuscany's evolution is especially notable because the Lorraine family viewed the small state as a laboratory for experimentation with Enlightenment ideals. Notably, Tuscany became the first state in the world to abolish the death penalty, among many other social and philosophical innovations. Pisa acquired distinct significance as a cultural epicenter during this period, while Florence, despite remaining the governmental and administrative center and the capital of the state, was still heavily influenced by the Baroque culture of the Medici era (Melis & Melis, 1996).

This view on Tuscany and Pisa underscores the essential role that architectural spaces dedicated to the arts played in the broader context of social and political reform. The development and transformation of theaters and other performative spaces not only mirrored the changing aesthetic tastes and cultural priorities of the time but also actively contributed to the spread of Enlightenment values and the nurturing of a public sphere where new ideas could be disseminated and debated. The Lorraine family's patronage of such projects in Pisa and across Tuscany exemplifies the connection between architecture, politics, and philosophy that marked this period of experimentation and innovation.

The Ephemeral Architecture of the Regency Years

The Regency Years (1737–65) marked the onset of this new era. The transition from the Medici, following the death of the last heir Gian Gastone, to the Austrians was quite complex due to the Treaty of Vienna: The Duchy of Parma and Piacenza was annexed to Austria, which, in turn, had to cede to Charles, the son of Philip V of Spain and Elizabeth Farnese, the Kingdoms of Naples and Sicily. After Tuscany was assigned to the Augsburg-Lorraine in 1737, Francesco, husband of Maria Theresa, the future Empress, granted full powers to Prince Marco di Craon.

Francesco's arrival in Florence two years later, in 1739, was an opportunity for several performative arts initiatives aimed at celebrating power, in line with traditions established during the Medici rule. The concept of ephemeral installations and artistic performances related to the celebration of power is a hallmark of the Baroque age. Similarly, in other cities, efforts were made to welcome the Grand Duke with solemn celebrations. For instance, an ephemeral structure was erected in Pisa's Piazza Garibaldi to welcome the sovereigns with arches, festoons, and friezes that camouflaged the old tower located where today stands the Casino dei Nobili (the House of the Nobles), at the very center of the city in front of the Arno River.

The scenography and decoration represent a parallel trajectory in architecture that would be interesting to trace throughout the eighteenth century, related to the celebration of sovereigns and the traditional taste for festivals and city events such as the "Gioco del Ponte."

The "Luminara," for instance, is an architectural light performance along the facades of the buildings, especially in the Lungarni. Goldoni described it as follows: "I saw one evening … the city all ablaze with an infinite number of small flames, which not only illuminated from top to bottom the houses, palaces, bridges, buildings, perspectives but also with beautiful arches, with artful machines, changed the city's aspect in such a way that I certainly cannot describe."

The emphasis on scenography and decoration during these events, as a broader trend within the architecture and urban planning of the era, reflects an understanding of urban spaces not just as sites for ordinary activities but as canvases for the expression of power, culture, and collective identity. Through the strategic use of ephemeral installations and performances, Pisa and other Tuscan cities were able to temporarily transform their public spaces into stages for the enactment of social and political narratives, thereby reinforcing the authority of the ruling class and celebrating the achievements of the Enlightenment.

Theatre as a Meeting Place Between Arts and Enlightenment

The transformation of spaces for the performative arts in Tuscany, with Pisa as a focal point, not only illuminates the architectural and urbanistic developments of the time but also reveals the broader socio-political dynamics at play. The legacy of these transformations, encapsulated in the remaining structures and in the cultural memory of the city, continues to influence Pisa's identity as a cultural epicenter, illustrating the lasting impact of the Enlightenment's ideals on the fabric of Tuscan society.

In 1766, for the arrival of Grand Duke Peter Leopold, extensive theatrical preparations once again transformed the city into a scenographic space for celebration, underscoring the significant role of architecture and urban decorations in legitimizing the power of the ruling class and weaving art into the fabric of political and social life.

On this occasion, the public theater in the Gambacorti Palace was adorned with ephemeral decorations, including mirrors and illuminated fans strategically placed throughout the boxes, complemented by white and red silk flags on the exterior. The first order was further embellished with flags of dried flowers in the same colors, subtly alluding to the attire of the Knights of St. Stephen. The box featured red flags patterned with silver, and the corners and center were enhanced with luminous busts depicting armed men and other trophies. The ceiling, supported by four large pillars "colored like lapis lazuli with golden bases and capitals," was artistically rendered by the renowned artists Giovanni Tempesti and Mattia Tarocchi (cit, in Melis & Melis, 1996).

The bond between the city of Pisa and the Lorraine Court was strengthened with the idea of constructing a grand ducal seat in Pisa, leading to the restructuring of the Grand Ducal Palace and the entire Piazza S. Niccola area. Three years prior, a low and extended structure, tentatively attributed to Pellegrini and likely coinciding with the Palace of the Finance Intendancy dated "1774," had been constructed in the square. Architecturally similar to the Grand Ducal Palace, this structure's symmetry and central entrance blended seamlessly into the uniformity of the architectural elements. During this period, the church of S. Niccola also underwent radical transformations, with its rear facade revealing unmistakable signs of 18th-century restorations, and the new Prini Theater (now Rossi) was created in 1770 as a central feature of the urban regeneration.

The Prini Theater, had as its promoter and architect "the chief master mason Orazio Cecconi," assisted by engineer Francesco Bombicci; they erected the building "where the garden of the old Dispensary was located, graciously granted by Grand Duke Leopold I, with the view of improving the country's fortune, because the old theater owned by the Municipality was no longer suited," already from '65, to contain its patrons and the numerous foreigners.

In this political phase, theaters were intended both for their educational capacity and as places where social classes could converge—a constant aspiration of reformist and Enlightenment thought—thus likening them to service architectures. However, Pietro Leopoldo's attitude towards theater was contradictory, as seen from 1785, with increased strictness towards some theater operators and directives to reduce the number of performance venues, signaling a shift in political aspirations influenced by the moralistic resurgence of the Jansenists and the differing political orientation of the Emperor, which hindered the completion of a Constitution.

In 1838, Ranieri Grassi's Description provides insights into the original structure of the theater and the 19th-century renovations, likely conducted by Alessandro Gherardesca, which made the building "quite elegant and harmonious," countering earlier criticisms. In 1798, Giuliano Francesco Prini transferred the property to a society of citizens called Accademia dei Costanti. After this society dissolved in 1822, another society called the Ravvivati took possession of the building in 1824 and, until 1858, executed works that gave the theater its definitive structure, which Grassi described in the mid-nineteenth century as "horseshoe-shaped" with 81 boxes or loggias arranged in four orders, including a large central box for the Court.

Consistent with the Lorraine Government's emphasis on sobriety, monumentality was eschewed in this construction in favor of a scheme common in Florentine theaters built during the Regency. This scheme is characterized by the "non-correspondence between the internal disposition and the external shell of the building," meaning the facade did not reflect the interior's use and articulation, resembling a more conventional residential building instead. This choice stemmed from the idea of the theater as a private institution. Beyond the functional aspects, the duality of this architecture—sobriety and essentiality on the exterior contrasted with an interior full of "surprises"—became a leitmotif of extra-urban Pisan architecture during the Lorraine period.

The various architectural and cultural transformations within Pisa during the eighteenth century highlight the relationships between social, educational, and political changes initiated by the Lorraine, particularly by Peter Leopold, in the suppression of noble and church prerogatives to convert them into public facilities dedicated to health and science, such as the anatomical theater, and to knowledge, like the scientific exhibitions in museums.

Similarly, to what happen in the old theater for the arrival of Francesco, the lavish decorations for the Prini theater, for the arrival of the Ferdinand III, the successor of Peter Leopold, in 1791, showcased the era's penchant for ornate and symbolic design, blending artistry with festivity to honor the monarch.

An anonymous author of the Description of the City of Pisa describes the building, in 1792, as "moderately spacious" and adds nothing more, believing it does not overly excite the interest of the "Traveler," who had seen "many others superior in Italy." This might explain why the theater sought a specific intervention to design a setting that met the court's expectations.

The interior of the theater was elaborately decorated to host a ball in honor of Ferdinand III's arrival: "The stage box was adorned with mirrors and ornamented with gold tinsel roses, painted and braided pillars. The room was draped in green and sky-blue fabrics, adorned with branches of flowers." (cited in Melis & Melis, 1996).

This interest in ephemeral architecture, with its ancient tradition, was renewed in the nineteenth century; it suffices to recall the magnificent setups for Piazza Santa Caterina, an excellent outdoor theatrical venue for celebratory representations echoing themes of Roman architecture. In this regard, it is worth mentioning that as early as 1739, in Florence, there was an occasion that rendered the celebrations for the arrival of the Grand Duke of specific interest: the construction of a triumphal arch in masonry by architect Jean Nicolas Jadot. This was not ephemeral architecture but rather the prelude to the "resumption" of a monumental theme.

In addition to focusing on the requalification of the territory in all its different environmental and architectural aspects, the attention of Francesco III's administrators turned towards promoting cultural services even in smaller centers. This indicates a shift in governance from the Medici Administration, which was still very much oriented towards an aristocratic idea of education, towards the Enlightenment era. In this phase, theaters began to be considered less as spaces for leisure and more as part of cultural and educational advancement where the public role becomes even more relevant.

In Tuscany, unlike other parts of the peninsula, the contrasts between radical reformers and rulers were rather mild, and, indeed, a close link was established between the Administration and the University of Pisa. Almost all of the most eminent ministers and officials of the Lorraine government had in fact studied in Pisa, which "was the only obstacle that prevented Tuscany from being reduced to the state of ignorance that almost all the rest of Italy was suffocated by."

From Grassi, we learn that "under the auspices of Ferdinand III, the economy of this pious establishment was regulated and improved; and recently," the Pisan author continues, "thanks to the generosity of reigning Leopold II, provisions were made for its greater healthiness by expanding the large male infirmary (separated by a long courtyard from the female one), opening ventilations on every side, redoing the flooring, etc.; with the addition of another space for Anatomy preparations, a dedicated room for maternity cases, another for those suffering from shameful diseases, new manna baths, and a splendid anatomical theater."

In 1836, "by the royal generosity of reigning Leopold II, the necessary sum was granted to unite the two divided portions of the museum," and to construct a large and well-designed amphitheater room, where botany and natural history lessons were to be held. By 1838, these works had been completed; the six rooms obtained from this last expansion of the Museum were allocated for the exhibition of comparative anatomy collections, invertebrate animals, and mineralogical collections.

This period indeed marked a key phase in Italian history with the emergence of the Risorgimento movements of the '20 s and '30 s, which harbored the will for independence of those peoples who felt oppressed by foreigners. Although this was not the case for Tuscans, the fruits of the revolutions of the late eighteenth century and the imposed restorations did unfold, albeit in a less dramatic manner.

The introduction of new urban capitalism was not an obstacle to the completion of large public works, as bourgeois private initiative became supportive of the new regeneration projects. The premises for urban renewal were primarily the persistent, and now urgent, need for decorum, the transformation of the city in a bourgeois sense with the construction and restoration of theaters, the erection of honorary monuments, and urban furnishings that conventionally integrated the arrangement of squares and avenues for strolling, according to the typical 19th-century taste for promenades.

The proliferation of theaters was again an example of the new social and cultural course of the bourgeois age. This renewed interest in Pisa was reflected in the restoration, in '24, of the Rossi theater, the construction in 1806 of the public open known as "arena Federighi", and its restoration in 1831 by Gherardesca, who on that occasion endowed it with boxes, staircases, and terraces. Only in the second half of the century did the construction of a new Politeama begin, another significant addition to the public theaters in Pisa, while discussions about a new theater and decisive interventions on the existing 18th-century structure were only initiated in the 1860s.

The unification of Italy marked a new phase for theater construction in the city. The new Teatro Verdi in Pisa represented a pivotal moment in the architectural and cultural development of the city. In contrast to the external sobriety of the old theater, typical of the 18th-century Lorraine style, the expressiveness of Teatro Verdi, though modest compared to contemporary examples, signified a revolution, albeit delayed, reflecting the successful typology pioneered by Milan's Teatro "La Scala," designed by Piermarini in 1778.

Teatro Verdi, initially called "Teatro Nuovo" and inaugurated in 1867, was spearheaded by engineer Ranieri Simonelli on behalf of a citizen's society. The design was attributed to the Florentine architect Andrea Scala. Funded by the municipality, Teatro Verdi stood out for its acoustic excellence and spacious architecture, accommodating approximately 1200 spectators. The theater, from its inception, was among the largest in Italy and abroad, reflecting Pisa's vibrant cultural and social aspirations of the era.

This theater, with its classical appearance, embodies the sensitivities of the New Era, where the enjoyment of performative arts is purely public and serves as a social engine of equity in which all societal layers participate. Symbolically, the external and internal architectural components communicate the values of the state, represented by the people of a bourgeois society, independent from foreign dominations.

The interior of the theater was elaborately decorated to host a ball in honor of Ferdinand III's arrival: "The stage box was adorned with mirrors and ornamented with gold tinsel roses, painted and braided pillars. The room was draped in green and sky-blue fabrics, adorned with branches of flowers" (cited in Melis & Melis, 1996). This interest in ephemeral architecture, rooted in ancient traditions, was revived in the nineteenth century, as exemplified by the magnificent setups for Piazza Santa Caterina, an outstanding outdoor theatrical venue that echoed themes of Roman architecture for celebratory representations.

In this regard, it is worth mentioning that as early as 1739, in Florence, there was an occasion that rendered the celebrations for the arrival of the Grand Duke of specific interest: the construction of a triumphal arch in masonry by architect Jean Nicolas Jadot. This was not ephemeral architecture but rather the prelude to the "resumption" of a monumental theme. In addition to focusing on the requalification of the territory in all its different environmental and architectural aspects, the attention of Francesco III's administrators turned towards promoting cultural services even in smaller centers. This indicates a shift in governance from the Medici Administration, which was still very much oriented towards an aristocratic idea of education, towards the Enlightenment era.

In this phase, theaters were increasingly viewed not merely as spaces for leisure but as integral to cultural and educational advancement, with a heightened emphasis on their public role. In Tuscany, unlike other parts of the peninsula, the contrasts between radical reformers and rulers were relatively mild, and a close link was established between the Administration and the University of Pisa. Almost all of the most eminent ministers and officials of the Lorraine government had studied in Pisa, which "was the only obstacle that prevented Tuscany from being reduced to the state of ignorance that almost all the rest of Italy was suffocated by" (cit in Melis & Melis, 1996).

From Grassi, we learn that "under the auspices of Ferdinand III, the economy of this pious establishment was regulated and improved; and recently," the Pisan author continues, "thanks to the generosity of reigning Leopold II, provisions were made for its greater healthiness by expanding the large male infirmary (separated by a long courtyard from the female one), opening ventilations on every side, redoing the flooring, etc.; with the addition of another space for Anatomy preparations, a dedicated room for maternity cases, another for those suffering from shameful diseases, new manna baths, and a splendid anatomical theater."

In 1836, "by the royal generosity of reigning Leopold II, the necessary sum was granted to unite the two divided portions of the museum," and to construct a large and well-designed amphitheater room, where botany and natural history lessons were to be held. By 1838, these works had been completed; the six rooms obtained from this last expansion of the Museum were allocated for the exhibition of comparative anatomy collections, invertebrate animals, and mineralogical collections.

This period indeed marked a key phase in Italian history with the emergence of the Risorgimento movements of the '20 s and '30 s, which harbored the will for independence of those peoples who felt oppressed by foreigners. Although this was not the case for Tuscans, the fruits of the revolutions of the late eighteenth century and the imposed restorations did unfold, albeit in a less dramatic manner.

The introduction of new urban capitalism was not an obstacle to the completion of large public works, as bourgeois private initiative became supportive of the new regeneration projects. The premises for urban renewal were primarily the persistent, and now urgent, need for decorum, the transformation of the city in a bourgeois sense with the construction and restoration of theaters, the erection of honorary monuments, and urban furnishings that conventionally integrated the arrangement of squares and avenues for strolling, according to the typical 19th-century taste for promenades.

The proliferation of theaters was again an example of the new social and cultural course of the bourgeois age. This renewed interest in Pisa was reflected in the restoration, in '24, of the Rossi theater, the construction in 1806 of the public open known as "arena Federighi", and its restoration in 1831 by Gherardesca, who on that occasion endowed it with boxes, staircases, and terraces. Only in the second half of the century did the construction of a new Politeama begin, another significant addition to the public theaters in Pisa, while discussions about a new theater and decisive interventions on the existing 18th-century structure were only initiated in the 1860s.

The unification of Italy marked a new phase for theater construction in the city. The new Teatro Verdi in Pisa represented a pivotal moment in the architectural and cultural development of the city. In contrast to the external sobriety of the old theater, typical of the 18th-century Lorraine style, the expressiveness of Teatro Verdi, though modest compared to contemporary examples, signified a revolution, albeit delayed, reflecting the successful typology pioneered by Milan's Teatro "La Scala," designed by Piermarini in 1778.

Teatro Verdi, initially called "Teatro Nuovo" and inaugurated in 1867, was spearheaded by engineer Ranieri Simonelli on behalf of a citizen's society. The design was attributed to the Florentine architect Andrea Scala. Funded by the municipality, Teatro Verdi stood out for its acoustic excellence and spacious architecture, accommodating approximately 1200 spectators. The theater, from its inception, was among the largest in Italy and abroad, reflecting Pisa's vibrant cultural and social aspirations of the era.

This theater, with its classical appearance, embodies the sensitivities of the New Era, where the enjoyment of performative arts is purely public and serves as a social engine of equity in which all societal layers participate. Symbolically, the external and internal architectural components communicate the values of the state, represented by the people of a bourgeois society, independent from foreign dominations.

The emphasis on scenography and decoration during these events, as a broader trend within the architecture and urban planning of the era, reflects an understanding of urban spaces not just as sites for ordinary activities but as canvases for the expression of power, culture, and collective identity. Through the strategic use of ephemeral installations and performances, Pisa and other Tuscan cities were able to temporarily transform their public spaces into stages for the enactment of social and political narratives, thereby reinforcing the authority of the ruling class and celebrating the achievements of the Enlightenment.

The various architectural and cultural transformations within Pisa during the eighteenth century highlight the interplay between social, educational, and political changes initiated by the Lorraine, particularly by Peter Leopold, in the suppression of noble and church prerogatives to convert them into public facilities dedicated to health and science, such as the anatomical theater, and to knowledge, like the scientific exhibitions in museums.

Similarly, to what happen in the old theater for the arrival of Francesco, the lavish decorations for the Prini theater, for the arrival of the Ferdinand III, the successor of Peter Leopold, in 1791, showcased the era's penchant for ornate and symbolic design, blending artistry with festivity to honor the monarch. An anonymous author of the Description of the City of Pisa describes the building, in 1792, as "moderately spacious" and adds nothing more, believing it does not overly excite the interest of the "Traveler," who had seen "many others superior in Italy." This might explain why the theater sought a specific intervention to design a setting that met the court's expectations.

Reference

Melis, A., & Melis, G. L. (1996). Architettura pisana: dal Granducato lorenese all'Unità d'Italia. Edizioni ETS

Chapter 10
The Stadium as a Performative Space

Alessandro Melis and Marco Pelle

Abstract *The Stadium as a Performative Space* examines the architectural and cultural significance of stadiums as enduring and transformative spaces within urban environments. The adaptability of stadiums like the Verona Arena and Arles Amphitheatre demonstrates the potential of these structures to meet evolving civic needs. Today, stadiums face challenges tied to single-use design and environmental concerns. However, modern innovations reimagine stadiums as multifunctional urban hubs and renewable energy sources. This chapter advocates stadiums' versatility, arguing that adaptive reuse can transform them into dynamic assets within contemporary cities, aligning architecture with environmental and social resilience.

Keywords Stadium as performative space · Architectural significance · Cultural importance · Urban environments · Verona arena · Arles Amphitheatre · Adaptability · Civic needs · Single-use design · Environmental concerns · Modern innovations · Multifunctional urban hubs · Renewable energy sources · Versatility · Adaptive reuse · Dynamic assets · Contemporary cities · Architectural alignment · Environmental resilience · Social resilience · Transformative spaces · Enduring structures · Urban development · Community engagement · Sustainability in design · Cultural evolution · Infrastructure challenges · Performative architecture · Civic identity · Innovative solutions

Historically, sports can be regarded as a type of performative art, thus placing stadiums within the context of performative art spaces offers a unique perspective. Stadiums have long been integral to urban architecture, with origins tracing back to Olympia and Rome. These structures were central to Western cities long before the emergence of medieval cathedrals and the industrial revolution's railway stations.

A. Melis (✉)
School of Architecture and Design, New York Institute of Technology, New York, USA
e-mail: amelis@nyit.edu

M. Pelle
Department of Social and Political Sciences, Bocconi University, Milano, Italy

A. Melis and M. Pelle (eds.), *Design for Performative Arts Spaces*,
https://doi.org/10.1007/978-3-031-98215-6_10

In contemporary times, however, stadiums face increasing scrutiny. Yet, this does not have to be their fate. Historical evidence demonstrates that stadiums can drive urban growth and adapt to the cultural needs of each era. Today, architects and planners are innovating ways to transform the once single-purpose sports arenas emblematic of twentieth-century modernity.

The amphitheatre in Arles, capable of hosting 25,000 spectators, exemplifies the versatility of stadiums. Constructed by the Romans in 90 AD, it transformed into a fortress with four towers after the fifth century, and later into a village with over 200 houses. In the nineteenth century, growing interest in conservation saw it revert to an arena for bull racing.

Similarly, Verona's arena, built 60 years before Arles' amphitheatre and 40 years before the Colosseum, has stood the test of time. Today, it is celebrated as a premier opera venue due to its exceptional acoustics.

Research by Taisuke Kuroda from Kanto Gaukin University highlights another fascinating example in Lucca. The Piazza dell'Anfiteatro, originally accommodating 10,000 spectators, gradually integrated into the city's fabric, serving various roles from housing to a market square, while retaining elements of its original structure.

Another prime example is the Pantheon in Rome, originally constructed as a temple to all the gods, which was later converted into a Christian church. This adaptability has allowed it to remain a vital part of the city's architectural and cultural landscape for centuries. The Teatro Marcello, also in Rome, transitioned from a theater to a fortress, then into a residential building, illustrating the inherent value of multifunctional architecture.

Modern stadiums share many features with ancient amphitheatres intended for public games, yet lost some flexibility with twentieth-century advancements in steel, reinforced concrete, and night-time lighting.

Many contemporary stadiums, located in suburban areas, are designed solely for sports and surrounded by large concrete parking lots, making them less accessible and more energy-intensive, contributing to urban heat.

However, architects like Herzog & De Meuron, Zaha Hadid Architects, and Toyo Ito are reimagining stadiums as urban hubs and power plants. The trend is toward mixed-use developments, incorporating public spaces and services beyond sports, such as hotels, retail outlets, conference centers, restaurants, playgrounds, and green spaces. This approach enhances land use efficiency and urban regeneration, drawing in a broader audience.

Examples in the UK include Wembley and Old Trafford, whose mixed-use facilities have become models for stadiums worldwide. The upcoming Fulham FC Riverside Stand Redevelopment aims to extend the riverside boardwalk, adding housing and retail.

The concept of stadiums as power stations arises from the need to address energy issues through interconnected buildings using smart grids. Stadiums, with their large canopies and significant height, are ideal for photovoltaic panels and micro wind turbines.

Freiburg Mage Solar Stadium represents a new generation of stadiums as power generators, alongside the Amsterdam Arena and Taiwan's Kaohsiung National

Stadium. Inaugurated in 2009, Kaohsiung's stadium features 8844 photovoltaic panels producing up to 1.14 GWh of electricity annually, reducing CO_2 emissions by 660 tons and supplying up to 80% of the surrounding area's energy needs when not in use.

Stadiums have continually adapted, serving as military garrisons, residential villages, public spaces, theaters, and now as centers for advanced engineering. Rather than becoming obsolete, modern stadiums integrate multiple functions to contribute to a sustainable urban future.

Focusing on a single use for a structure, akin to cherry-picking, overlooks the broader potential and adaptability of architecture. This is why stylistic restorations that emphasize a particular style or use from the nineteenth century are problematic, especially in times of environmental and social crises. Embracing the versatility of architecture ensures resilience and relevance across different eras and needs, making it a crucial asset in contemporary urban planning.

Reference

Melis, A. (2019). Stadiums aren't fated to disrepair and disuse—history shows they can change with the city. The Conversations.

Chapter 11
Performative Art at the New York Tech School of Architecture and Design: From Education to Practice (2021–2024)

Maria Perbellini, Giovanni Santamaria, Marcella Del Signore, Sung Park, Gertrudis Brens, and Alessandro Melis

Abstract *Performative Art at New York Tech's School of Architecture and Design* explores the intersection of architecture and performative art through three innovative projects, embodying the creative, fluid nature of design education. This chapter details how New York Tech SoAD has redefined traditional architectural exhibitions into interactive experiences that engage participants directly, showcasing a transformative approach to architecture through process-oriented exhibits and multidisciplinary collaboration. Each project exemplifies adaptability, merging architectural practice with sustainability and social engagement. The Italian Pavilion at the 2021 Venice Biennale pioneered exaptive design, adapting spaces for resilience. The 2023 "Students as Researchers" Biennale exhibit highlighted academic research as an agent of environmental and social change, while the "Future by Design" showcase at Milan's Salone del Mobile demonstrated sustainable, interactive designs on various scales. Together, these case studies underscore SoAD's commitment to rethinking architecture, advancing its capacity to foster resilient, adaptable communities for a changing world.

Keywords Performative art · New York tech · School of architecture and design · Architecture and art intersection · Innovative projects · Design education · Architectural exhibitions · Interactive experiences · Transformative approach · Process-oriented exhibits · Multidisciplinary collaboration · Adaptability · Sustainability · Social engagement · Italian Pavilionm · 2021 Venice Biennale · Exaptive design · Resilience · "Students as researchers" exhibit · Environmental change · Social change · "Future by design" showcase · Milan's Salone del mobile · Sustainable designs · Interactive designs · Case studies · Rethinking architecture · Resilient communities · Adaptable communities · Changing world

---

M. Perbellini · G. Santamaria (✉) · M. Del Signore · S. Park · G. Brens · A. Melis
New York Institute of Technology, New York, USA
e-mail: gsantama@nyit.edu

A. Melis and M. Pelle (eds.), *Design for Performative Arts Spaces*,
https://doi.org/10.1007/978-3-031-98215-6_11

Introduction to Performative Art in Architectural Practice

This chapter investigates the intersection of performative art and architecture through three distinctive case studies promoted by the New York Tech School of Architecture and Design. These case studies transcend traditional exhibitions, embodying the essence of performative art due to their deeply process-oriented nature. Unlike static displays, they integrate the creative process directly into the exhibited work through videos, interactive devices, and other multimedia elements.

The human agency involved before and during the exhibition is as crucial as the exhibition itself. This approach underscores the notion that education in related fields of creativity is inherently performative. The visitors are not mere observers but active participants in the art exhibition, engaging in a continuous, iterative process where the authors are also participants. This dynamic engagement fosters an environment where unexpected outcomes are not just anticipated but embraced, highlighting the fluid and interactive nature of architectural practice.

The case studies are designed to challenge conventional boundaries and encourage spontaneous, creative responses. They illustrate how performative art in architecture can create spaces that are not only functional but also responsive to the needs and interactions of their users. By involving the audience directly in the creative process, these projects emphasize the evolving relationship between space, creator, and participant, and how each influences the final outcome.

In this context, the performative aspect of these architectural projects lies in their ability to adapt and evolve through interaction. The case studies demonstrate how architectural spaces can serve as platforms for ongoing dialogue and creative exchange, pushing the boundaries of traditional architectural practice and education.

Case Study 1: Resilient Communities at the Venice Biennale 2021

The "Resilient Communities" project for the Italian Pavilion at the Venice Biennale 2021 stands as a quintessential example of exaptive design in architecture. Exaptation, a concept derived from evolutionary biology, refers to the repurposing of structures with latent potentials for new, unforeseen functions (Pievani, 2024). In the architectural realm, this translates to the creative transformation of existing structures without a predetermined final purpose, allowing for novel and innovative uses.

The design for the Italian Pavilion was inspired by the spontaneous emergence of new uses for existing structures. It integrated principles of environmental graphic design and cutting-edge technologies to dynamically repurpose urban spaces and enhance resilience. By embracing diversity, variability, and redundancy, the project illustrated how exaptive design could effectively address environmental crises and societal needs (Melis, 2021a, 2021b).

Celebrating architectural disorder and structural plasticity, the pavilion emphasized the benefits of non-adaptive mechanisms found in nature. It championed designs that adapt fluidly to changing environments, countering the rigidity of traditional planning frameworks. The pavilion's radical and disruptive technologies were designed to continually reassess and redefine architectural concepts, promoting more economical and functional design approaches (Melis, 2021d, 2021e).

By incorporating non-deterministic and collective approaches, the pavilion fostered an environment of creativity and intimacy. This challenged conventional architectural perceptions, advocating for designs that embrace uncertainty and adaptability. The pavilion's physical organization allowed for a dynamic and informal visitor experience, critiquing the limitations of rigid exhibition models. It emphasized ecological sustainability through its embrace of redundancy, variability, and diversity, illustrating a holistic approach to resilient and adaptable architectural design.

Case Study 2: Students as Researchers: Integrating Pedagogical Innovations

In 2023, the Venice Biennale featured "Students as Researchers: Creative Practice and University Education," curated by Maria Perbellini, Marcella Del Signore, Athina Papadopoulou, and Sandra Manninger. This exhibition was distinguished by its integration of innovative technologies from a pedagogical perspective, underscoring the pressing need for action across social, environmental, political, and economic domains.

As part of the eighteenth International Architecture Exhibition, the showcase included contributions from 20 universities. It emphasized creative practices and processes nurtured by students and faculty to address urban challenges and climate issues. The projects underscored the significance of civic engagement, community well-being, and safety, all of which are integral to the curricula of the participating universities.

Projects spanned various scales and explored computational design tools, 3D printing technologies, and bioinspired materials. At the urban scale, resilient community models and sustainable growth strategies were prominently featured, incorporating principles of the circular economy. This exhibition highlighted the crucial role academic institutions play in fostering transformative approaches to design, thereby preparing students to tackle urgent social and environmental issues (Cousin, 2006; Del Signore & Riether, 2020; Melis, 2021c).

The research projects featured in this exhibition addressed themes across different cultures and educational models. At the material scale, projects delved into computational design tools and 3D printing technologies for bioinspired geometries and regenerative biobased materials. These investigations not only integrated new technologies but also reinterpreted indigenous and local crafts and techniques (Melis, 2021a, 2021b).

At the building scale, projects strategically utilized biobased resources to reduce carbon footprints and energy consumption, addressing accessibility and inclusion by challenging traditional architectural standards. At the urban scale, projects explored resilient community models that adaptively respond to climate change, incorporating principles of the circular economy and strategies for sustainable urban growth. Furthermore, by implementing artificial intelligence and machine learning algorithms, students uncovered hidden urban elements and questioned existing data structures and design methods.

Teaching can become a powerful research instrument through the exchange of critical ideas with students, facilitated by bidirectional pedagogical models where teachers and learners collaborate. Under the guidance of faculty and practitioners from various design disciplines, students could critically and experimentally interrogate existing modes of production, consumption, building, and living. "Students as Researchers" thus brought together research outputs from institutions worldwide, offering fresh perspectives to address pressing social and environmental issues and underscoring the responsibility of future designers (Melis, 2021d, 2021e).

Case Study 3: Future by Design at Salone Del Mobile Di Milano

The "Future by Design" exhibition at the SaloneSatellite during the Salone del Mobile di Milano represented the New York Tech School of Architecture and Design. Curated by Maria Perbellini and Giovanni Santamaria, the exhibition showcased collaborative and multidisciplinary work, emphasizing the integration of creativity, functionality, and sustainability across different scales.

Central to the exhibition was the "Entangled Matter Wall" (EMAW), which acted as a catalyst for creative moments and cross-disciplinary dialogues. This feature incorporated modular components, organic recyclable materials, and computational methodologies, fostering sustainable and adaptable designs. This installation highlighted the collective efforts of faculty and students, stressing the significance of diversity and environmental awareness in design (Melis, 2022).

The EMAW promoted proactive dialogues and interactions between curricular and extracurricular activities. It exemplified how sustainable practices and innovative integration could create adaptable and resilient structures. The exhibition underscored a dedication to reducing consumption and waste, demonstrating the ongoing journey of the installation from Venice to Milan and back to New York.

The collaborative and multidisciplinary efforts displayed within the New York Tech pavilion "Future by Design," curated by SoAD Dean Maria Perbellini and Associate Dean Giovanni Santamaria, also highlighted an important and enduring collaboration with the Town of Peccioli and the company Belvedere S.p.A., part of the Sistema Peccioli in Tuscany, Italy. These entities co-sponsored the project alongside New York Tech, IDC Foundation-NY, and the Alumni and Friends of the School of Architecture and Design. The design and fabrication team included Brian Polgar, Fadhil Fadhil, Elijah Williams, Ana Finkelstein, John Bermudez, Mario Medina Vilela, and Athina Papadopoulou, with graphic design by Marcella Del Signore, Evan Shieh, and Florencia Vetcher. The video editing team was composed of Kevin Park, Jake Mangan, and Nouman Wajid from the Department of Digital Art and Design, SoAD. Special recognition for the design was given to Marcella Del Signore, Cordula Roser Gray, Tatiana Teixeira, and Marcelo Chaves. The work involved contributions from students and faculty of undergraduate and graduate programs at the SoAD—New York Tech, including Bachelor of Architecture and Bachelor of Science in Architectural Technology with Chair Trudy Brens; Bachelor of Fine Arts in Interior Design with Director Florencia Vetcher; Master of Architecture with Interim Director Marcella Del Signore, who is also the Director of the Master of Science in Architecture, Urban Design; Master of Science in Architecture, Computational Technology with Director Alessandro Melis; Master of Science in Health and Design with Director Christian Pongratz; and Bachelor of Fine Arts in Digital Art & Design with Chair Kevin Park. The local installation was managed by WAY S.p.A., with local contractor Alessandro Cocciolo at G&C Contractor and local coordination by Dr. Arch. Michele Moreno.

This project investigated design processes and innovative integrations within the design field, balancing creativity, functionality, and sustainability across various scales, from materials to territories, while promoting inclusivity and technological advancement (Del Signore & Riether, 2020; Melis, 2021a, 2021b).

Employing an ecosystem and holistic approach, faculty, students, and researchers at the SoAD fostered diverse experimental computational methodologies and platforms for new transdisciplinary collaborations on applied research involving sensitive topics and design applications in environmentally vulnerable areas. This approach also evaluated the sustainable qualities of academic production while visualizing and coordinating their resiliency.

The Entangled Matter Wall (EMAW) was envisioned as a catalyst and common ground for several "creative moments" collected in a "meta-bookshelf." This installation, realized through the collaborative work of the Fabrication and Robotic Labs at SoAD, encouraged proactive dialogues and provided opportunities for fertile interactions between curricular and extracurricular activities to envision better and more balanced futures for all. The sustainable approach, aiming to reduce consumption and waste, drew inspiration from a previous installation for the Biennale di Venezia titled "Entangled Matter" by CRGArchitecture with X. Topia. This installation was integrated into the new design, inspiring its geometries and construction methods, and incorporating organic recyclable components. The new design explored possibilities for recycled, homogeneous, and low carbon footprint materials that were easy to produce, assemble/disassemble, and composed of modular components that guaranteed transformability and adaptability to different locations, conceived as a continuously evolving installation. Careful evaluations of materials and methods of production, transportation, and relocations were crucial to the project, which continued its journey from Venice to Milan and then back to New York (Gregotti, 2014).

The structure of the Entangled Matter Wall was populated with models exploring design processes at various scales with different materials, including organic ones, technologies of production, and operational tools that were part of student and faculty research within academic programs. These models were placed within dedicated open pods that were continuous with the enclosed ones from the original structure, containing green microenvironments capable of self-regeneration and maintaining controlled climatic conditions. The students' work symbolically recreated explorations, enhancing the production of new potential environments made possible through creative and applied knowledge (Pareyson, 2002).

Conclusion

These case studies illustrate how the New York Tech School of Architecture and Design integrates performative art into architectural practice. By embracing exaptive design, innovative pedagogies, and multidisciplinary collaboration, these projects highlight the potential for architecture to address environmental challenges and societal needs creatively and sustainably. Through these efforts, the school continues to push the boundaries of traditional design, fostering a more resilient and adaptable built environment for the future.

References

Cousin, G. (2006). An introduction to threshold concepts. *Planet, 17*(1), 4–5.

Del Signore, M., & Riether, G. (2020). *Urban machines: Public space in a digital culture.* Routledge.

Gregotti, V. (2014). *Il Territorio dell'Architettura.* Feltrinelli.

Melis, A. (2021a). Community resilience through exaptation. *FORUM A+P, 22*, 70–77.

Melis, A. (2021b). *Periferia e Pregiudizio.* University Press.

Melis, A. (2021c). The ecological waste. In S. Massaro (ed.), *Circular living.* D Editore.

Melis, A. (2021d). Resilient Community. *AND, 39.*

Melis, A. (2021e). Decalogue of the contents of the Italian Pavilion. In A. Melis & T. Pievani (Eds.), *Architectural exaptation. Catalogo del padiglione Italia. La XVII Biennale Venezia (Vol. 1a).* D Editore.

Melis, A. (2022). The demiurgic position of architecture in times of environmental crises. In G. De Finis & C. Pecoraro (Eds.), *Periferi@.* Castelvecchi.

Pareyson, L. (2002). *Estetica: teoria della formatività.* Bompiani.

Pievani, T. (2024). *Imperfection.* MIT Press.

Chapter 12
INTRA-Warhol: A Retrospective Reflection on Body, Space, and Media Through a Practice-Based Explorative Work

Antonino Di Raimo

Abstract This chapter focuses on a retrospective reflection of the past and ongoing exploration of the relationship between body and space. Among many others, this year's long research was initiated by a project conceived and delivered at the Flux Laboratory in Geneva in 2007. The project was centered around the multidisciplinary production titled "Trans-Warhol". "Trans-Warhol" was primarily shaped through a choreographic piece staged by the resident choreographer, Nicolas Musin, along with the creative direction of Cynthia Odier at Flux. At that time, I was a GAI (Giovani Artisti Italiani) funded scholar, concurrently working on a practice-based experimental project that bridged architecture and dance. My sole constraint was to remain a participant observer of the work and to relate my experiment to the Flux Lab's current production. The project utilizes both practice and its exegesis to explore the relationship between the body and space, interpreted through the lens of Andy Warhol's work and his prophetic insights into the role of contemporary media in subsequent years.

Keywords Practice-based experimental project · Architecture and dance · Participant observer · Exegesis · Contemporary media · Trans-Warhol · Multidisciplinary production · Choreographic piece · Retrospective reflection · Body and space · Research exploration

In 2007, I was involved as a funded participant in a transdisciplinary project developed at the Flux Laboratory in Geneva. I held a grant that allowed me to act as a participant observer in the project for a couple of months while working on my own exploratory project between architecture and dance.

The Flux Laboratory had begun working on a rather ambitious interdisciplinary production: Trans-Warhol (Musin & Odier, 2007), which premiered in March 2007

A. Di Raimo (✉)
School of Architecture, Art and Design, University of Portsmouth, Portsmouth, UK
e-mail: antonino.diraimo@port.ac.uk

A. Melis and M. Pelle (eds.), *Design for Performative Arts Spaces*,
https://doi.org/10.1007/978-3-031-98215-6_12

on the stage of the Bâtiment des Forces Motrices in Geneva. Trans-Warhol was developed in collaboration with Kenneth Goldsmith, author of "All Our Mirror: The Selected Andy Warhol Interviews" (Goldsmith, 2004), from which the opera's libretto was inspired. With choreography of Nicolas Musin, the performance emerged as a contemporary opera-dance, underpinned by the relationship between the themes of brand, business, and art, along with their contradictions. Cynthia Odier, Director at the Flux, overseeing the whole project, advocated for tangible interdisciplinarity behind the production. The Flux-Laboratory decided to extend and finance my stay for another six months until the premiere, allowing me to inquire deeper into my exploratory project. On a daily basis, I had the opportunity to follow the work of the choreographer, the dancers, and the more general rehearsals of the performance. My work continued on a theme that is still central to my interests: the body-space relationships in architecture considered from a computational point of view. In the case of this project, the computational aspect gradually focused on the crucial aspect of the relationship between media and artistic production, so central in Warhol's work and already highlighted in several research works read at that time and after (Crone, 1987; Gopnik, 2021; Scherman & Dalton, 2009; Warhol, 1977).

While the construction of the opera-dance developed and involved various disciplines of artistic production, such as musical composition, the parts of the libretto inspired by Kenneth Goldsmith's work, and the choreographic work, I started to develop an exploration that was practice-based and oriented. I considered different aspects related to Warhol's artistic production. While focusing on them, I would retrospectively say that I adopted a meandering approach as a research method which would allow me to develop and share inquiry and ideas as much as it would happen within an educational setting (Friedman et al., 2020). I was keen to capture and catalyze different views, delivered by various agencies including the analytical work developed by scholars, which remained at the centre of diverse lines of inquiry and fundamental tools in building a narrative. This narrative would be later published in the form of a commentary for the Architetto Italiano magazine (Di Raimo, 2011). My argument was that different techniques implied by Warhol's work, such as silkscreen, xerox, oxidation, and later a computer Amiga, concerned two fundamental conceptual narratives: firstly, the use of procedures which implied a mechanical mode to produce artistic artifacts, and secondly, a protocol of the progressive disappearance of the artist's presence, which I would argue today was more about a protocol of propagation. Direct repetitions of works, or analogue clones with some variations, were a consequence of Warhol's way of working.

All these aspects, emerging in the famous statement 'The reason I'm painting this way is that I want to be a machine' (Swenson, 1963) are still relevant in our contemporary reality; and it is not coincidental that often Warhol's work has been reported in various ways as linked to contemporary production with respect to the role digital media.

Yet, as already intuited by Benjamin in his essay "The Work of Art in the Age of Its Technical Reproducibility" (Benjamin, 2008), the capability of the machine to allow reproduction became the centre of a new condition for creative production; much more than the creator.

I assume here, the machine and medium as two concepts both converging in the notion of apparatus, built by Agamben based on the work of Foucault: namely "literally anything that has in some way the capacity to capture, orient, determine, intercept, model, control, or secure the gestures, behaviours, opinions, or discourses of living beings" (Agamben, 2009). I will also focus on the Medium, in its status of constituting a middle condition, an agency that allows something to convey or accomplish something else. I will question the subjectivity and desubjectification implied by the notion of apparatus while assuming that its agency, namely the network of techniques between the individual and the world, is a tangible means for creative investigation.

My work at Flux started as an explorative practice (Badura, 2013) capable of blurring and dissolving my theoretical assumption in the practice and vice versa. I assumed that the creative exploratory practice implied a continuous iteration between theoretical discourses and individual tests. I also assumed that the creative work developed around me could trigger an exegesis of that practice research. I assume practice-based research in the sense described by Candy (2006) "Practice-based Research is an original investigation undertaken in order to gain new knowledge partly by means of practice and the outcomes of that practice" and that "contribution to knowledge may be demonstrated through creative outcomes in the form of designs, music, digital media, performances and exhibitions. Whilst the significance and context of the claims are described in words, a full understanding can only be obtained with direct reference to the outcomes".

However, working in parallel with the development of an interdisciplinary production that culminated in a performance encompassing dance, acting, opera singing, and music, continuously allowed me to observe and sometimes co-participate to other artistic practices. Each demonstrated its own agency although all were framed within the familiar protocol of the opera-dance. Yet, every practice I observed also aimed to develop its own creative contribution to the opera, serving as an exegesis of Warhol's work and its interplay with the business world. Ultimately, each practice constituted an exegesis of itself, allowing insights around the obstacles encountered in attempting to genuinely work within an interdisciplinary setting.

This type of circumstance also occurred in my explorations: an exciting discovery was Warhol's Dance Diagram, where a set of instructions is reproduced on a white canvas, similar to those used for learning ballroom dancing (Warhol, 1962). Essentially, while the diagram invites people to move by showing a set of instructions consisting of footprints, letters, numbers, and dotted arrows, it also remains fundamentally a painting. It is considered a sequence guiding the body in executing correct movements necessary for the completion of a specific step or dance style. This diagram would work well only if placed in an appropriate space, or better a medium in which the actions implied by the diagram would be easily triggered, as observed by other scholars (Gahr, 2015).

Indeed, the Dance Diagrams, more than just a documentation system like Labanotation, which records every minimal change in movements, celebrate the representation of a protocol; they embody the desire to represent the rules of an exchange, rather than the exchange itself. I started to consider them as an attempt to orchestrate an interaction between a hypothetical user and a space, with the painting acting as a medium within the medium of the space. The Dance Diagrams then overlap with some thought emerging from the dance rehearsals I occasionally attended in the corner of Flux's White ballet room.

While I was fascinated by the concise and powerful representation of the Dance Diagrams, I would compare it with the diagram as a traditional tool in architecture, useful for describing potential spaces or analyzing existing ones. The diagram, particularly popular in the first decade of the 2000s, also evoked the idea of a script—a rough draft to follow, which dictated protocols in anticipation of a performance. On the other hand, the work that the Flux Team was developing proceeded along a similar path. It was not uncommon to hear intriguing discussions about how the libretto had been developed or about the musical composition accompanying it. Sometimes, I could notice several discussions on the choreography, which unfolded according to a rather contemporary language, while some lamented the absence of any "pop" dimension. That intriguing conversation about the genesis of the opera, could have been represented through a complex diagram.

Certainly, my work did not follow any rigorous or analytical path, and ultimately Warhol was an opportunity to reflect on themes that interested me, primarily. I certainly did not claim to explain his work or reveal hidden aspects, although I would spend several hours viewing and reviewing "Andy Warhol Giant Size," an extra-large book of around 600 pages which contained mostly visual documentation of his work (Press et al., 2006).

Furthermore, I did not find that work aesthetically appealing. Rather, what I considered the persistent charm of it, something more intellectually than aesthetically attractive, was the pervasive two-dimensionality of those works, their ability to convey flatness, and a certain susceptibility to being reproduced in series, although through variations. Those works, ultimately always organized according to a grid, almost a two-dimensional or three-dimensional scaffolding-logic if necessary, like the Brillo boxes (Warhol, 1964). Certainly, in my eyes, there was no tangible corporeality in those works, or at least this was not the dimension that Warhol's works emphasized. In retrospect those works seem to me to be the representation and celebration of an apparatus of reproducibility and propagation. It was almost as if the works wanted to discuss the medium through which they were produced. The pervasive two-dimensionality implied a continuous reference to the screen rather than to the canvas.

My practice-based research at the time involved exploring three-dimensional representation in architecture through computer-assisted graphic techniques. Alongside generative work via algorithmic procedures, my ongoing interest has intensified in the meaningful absence of the body in computational research, particularly as generative techniques are now being supplanted by AI. This absence interests me both in terms of the representation of the human body and its agency throughout the design process and the experience of space.

The agency of the body was a topic frequently debated, often in reference to the seminal work of Pallasmaa (2005). However, I did not fully share Pallasmaa's criticism of the presence of computers in architecture design. Nor did I agree with those who, from the opposite perspective, predicted the emergence of a new, somewhat obscure architectural style driven by computational procedures (Schumacher, 2009).

My positionality lies somewhere in the middle and cannot surely rely on the adoption of a theorem or a singular epistemology. Coming from a background that also included training as a dancer, the theme of the body in architecture remained foundational to me.

My previous research partly conducted for master's degree thesis involved creating three-dimensional virtual dancers who, through animations, generated spatial geometric entities—almost always trajectories traced by selected parts of the body during animations. While I do not see myself strictly as an animator or a digital artist, this work allowed me to introduce an avatar, almost a virtual twin, into the two-dimensionality of the screen. This aspect of my work is documented in Di Raimo (2010).

In addressing the virtual dancer, I used the term "avatar" when it was not yet widely popular, though historically its adoption in computer science literature stemmed from video game culture and was later embraced by various online communities.

Looking back, my practice at that time, through the concept of an avatar, sought to reintroduce what I now define as human agency into three-dimensional space, mediated by the machine and its procedures.

However, in light of the work developed and discussed around me, it was difficult to discern any human agency in Warhol's works, especially in his paintings. While the Trans-Warhol opera-dance evolved into a very complex piece, I continued to ponder the Dance Diagrams. Yet, the dilemma about human agency, resembling a puzzle made up of various mediums, though dispersed, remained latent and only emerged through a few critical studies that highlighted biographical aspects of Warhol's life. Certainly, a common underlying notion was that of the icons, the devotional images used in Byzantine and Eastern Churches. Icons, through the medium of two-dimensional representation—a mere infinitesimal section of space—underscored the existence of a liminal border between a mystical world and the actual one (Dillenberger & Warhol, 1998; Livingstone, 1990).

Reviewing Warhol's photographic production in this light actually unfolds profound insights, considering the countless images often portraying friends and participants of that creative hub known as the Factory. Warhol's interest in Polaroid photography, which produced diverse snapshots, further reinforced this idea. Undoubtedly, he was a precursor of the pervasive use of contemporary media. My practice then, perhaps needed to focus on just one aspect: an exploratory and creative exercise centred on one of the iconic representations of his world.

Among hundreds of visual artworks, my attention was drawn to one of the photos at the Factory taken by Philippe Halsman entitled 'Andy Warhol's Factory' (Halsman, 1973). This photo presented two paramount aspects: (a) a small crowd of bodies grouped and positioned in a way that reminded me of the icons within an iconostasis; (b) rather than Warhol, it was Halsman who took the image, with Warhol himself seated and partially hidden within the small crowd. This image, therefore, unveils mutual gazes and representations of reciprocal mirroring, conditions which I would

Fig. 12.1 Reverse of the photograph taken by Philippe Halsman entitled 'Andy Warhol's Factory'—spotting and reconstructing human agencies alongside the three-dimensional space. *Author* Antonino Di Raimo

later consider emblematic of the presence of media in artistic production and their convergence in the notion of apparatus.

Since Warhol did not take the image, I wondered how the setting and the poses of the participants were decided. Was Halsman playing Warhol? And how did the medium of the camera play into this situation? Who was the creator of this photo? The machine in Halsman's hands, or perhaps Warhol's suggestions?

I decided then that a creative and convincing practice would involve descending into the photograph, deconstructing and reconstructing it to reveal its salient spatial characteristics.

From the perspective of my practice, this would imply that rather than performing a perspective restitution or uselessly employing photogrammetry software (for which I would have needed more photos of the same scene), my approach would lead me to analyse and reconstruct a hypothetical three-dimensional scene starting from the two-dimensionality of the photograph, including the three-dimensional reconstruction of the bodies. From a spatial viewpoint, this practice aimed to restore the spatial arrangement of the bodies. The bodies therefore had to be reconstructed and extracted from the photograph like any other spatial agent (Fig. 12.1).

A digital version of the photograph was imported into a three-dimensional application, maintaining its aspect ratio. I adopted a standard height for the body (one metre and seventy centimetres), which implied the viewpoint of the individual who took the photograph. The 3D application also enabled me to use a module dedicated to the modelling and animation of human-like objects.

Analysing the photograph to understand the arrangement of the bodies—a process developed entirely in two dimensions—was the first fundamental step. This step allowed me to identify the bodies (18 in total) and analyse their positions in depth. Even though the analysis was two-dimensional, it enabled me to hypothesise about the varying depths of body displacement and gain a preliminary idea of their arrangement, as if on a stage. The analysis also revealed different heights for the bodies located furthest from the observer. The scene had been carefully planned and was certainly not the result of an improvised pose in a spontaneous manner.

The identification phase was followed by that relating to the poses. The body pose, created completely by hand although within a digital environment, was enhanced by the use of the bone systems available in the 3D application.

As soon as the actual poses were achieved, I could see how the bodies lent meaning to the space. They conveyed a specific body language, which, it could be said, revealed the space through the poses far better than any architectural drawing would have done (Alexander, 1999).

At the same time, and without relying on any algorithms, patiently positioning the bodies within the fabricated 3D space created reciprocal relationships between the bodies themselves. The small crowd was clearly articulated and arranged, forming an ovoid pattern when observed from a top-down perspective.

However, the bodies were diverse; they exhibited different typologies and the skeletal system was certainly not enough to render them recognisable. The second most significant step, therefore, was related to the reconstruction of those bodies. What I was looking for was not verisimilitude, but rather the idea of human agency, or a graphic representation of that concept, in a three-dimensional space. Using open-source software, which provided ready-made neutral meshes, and relying solely on the available documentation (the photo), the patient and progressive transformation of the mesh topologies appeared to further unfold what I was searching for. I only realise now how the idea of modelling the clothes never emerged in this process. In reality, more than the difficulty or the time it would take to model other details, I believe that the unique agency of the bodies, reduced to the essentials (excluding hair and genitals from the modelling as well), saturated the desire of other data on a visual level. Reflecting on it now, perhaps I was seeking a form of essentiality, a sort of intrinsic expressive capacity of human topology, to convey corporeality in space.

By modelling individual bodies based on the photograph, the level of observation also increased. While everyone was looking directly or indirectly at the camera, only one character had his gaze fixed upwards, Warhol himself. This was the only singularity in a photograph that otherwise appeared to be constructed entirely according to the principles of central perspective. This process of three-dimensional restitution of an almost flat image, close to the representation of icons, raised reflections on the nature of perspective and orthogonal projections. Particularly, it brought to mind

Pavel Florenskij's work on reverse perspective (Florenskij, 2020). Florenskij's study raises questions not so much about reversed or "straight" perspective, but, I believe, about the perspective point of view that is assumed in the presentation, or about the co-presence of different viewpoints within the same image. Following this line of thought, instead of using an abstract camera, I quickly modelled a body, (present only in the three-dimensional digital drawings), which I considered to be a sort of projection of the photographer into the scene: a virtual alter-ego of Halsman.

Having a virtual camera, associated with the virtual body representing the person who had photographed the scene, employed considering the gaze belonging to each individual body. The scene began to show different points of view, although articulated within a purely virtual and three-dimensional environment, which raised questions about representation itself. Furthermore, having the bodies as a whole and their potential points of view opened up an unprecedented scenario, where the representation of one medium (the photographer and his camera) within another medium (the three-dimensional application) seemed to establish a world that exhibited properties of autonomy.

While the production of the Trans-Warhol was approaching its premiere, continuing to explore his work, I discovered that not only had Warhol explored the first computer graphics through an Amiga computer (Szabó, 2018), but that shortly before his death he had also planned the use of a robot in collaboration with Álvaro Villa Galvis, a sort of replica of himself that would interact with the viewers of a television show called the No Man Show.

As argued by Otty (2005), the emphasis on telepresence, or the relationship with the media space culminating as a buffer zone, would emerge in erasing Warhol from the process. A de-subjectivisation in the sense of Agamben?

My practice suggested the opposite. Rather than one of cancellation, or even de-subjectivisation, the space emerging from the practice showed me a place to explore, where the author would be projected and reflected in a potential game of mirrors, with the apparatus emphasising his presence, providing different versions of his agency and perhaps dramatically increasing his possible negotiations. What claims to be unique.

Acknowledgements I would like to thank Mme Cynthia Odier, Director at Flux Laboratory, Geneva, for her immense support during my time there between 2006 and 2007.

References

Agamben, G. (2009). *"What is an apparatus?" and other essays: And other essays*. Stanford University Press.

Alexander, M. D. (1999). *Body language*. The Museum of Modern Art.

Badura, J. (2013). Explorative practices in dialogue: Art-based research at the interface of arts, sciences, and design. In B. Imhof et al. (Eds.), *What is the Architect Doing in the Jungle? Biornametics*. Springer.

Benjamin, W. (2008). *The work of art in the age of mechanical reproduction*. Penguin, UK.

Candy, L. (2006). *Practice based research: A guide*. Creativity & Cognition Studios, University of Technology.
Crone, R. (1987). *Andy Warhol: A picture show by the artist, the early works 1942–1962*. Prestel-Verlag.
Dillenberger, J., & Warhol, A. (1998). *The religious art of Andy Warhol*. Burns & Oates.
Di Raimo, A. (2010). Architecture as caregiver: Human body—Information—Cognition. In: *LIFE in, On Responsive Information and Variations in Architecture: Proceedings of the 30th Annual Conference of the Association for Computer Aided Design in Architecture (ACADIA)* (pp. 110–116). Cooper Union, Pratt Institute. ISBN 978-1-4507-3471-4
Di Raimo, A. (2011). Warhol: Lo Sguardo della Macchina. In A. Saggio (Ed.), *Architettura & information technology* (pp. 88–96). Mancosu Editore.
Florenskij, P. A. (2020). *La prospettiva rovesciata* (2nd ed.). Adelphi.
Friedman, V. J., Robinson, S., Egan, M., Jones, D. R., Rhew, N. D., & Sama, L. M. (2020). Meandering as method for conversational learning and collaborative inquiry. *Journal of Management Education, 44*(5), 635–650. https://doi.org/10.1177/1052562920934151
Gahr, S. (2015, July 31). *The art of dance*. The Andy Warhol Museum. Retrieved from https://www.warhol.org/the-art-of-dance/
Goldsmith, K. (2004). *I'll be your mirror: The selected Andy Warhol interviews 1962–1987*. Carroll & Graf.
Gopnik, B. (2021). *Warhol: A life as art*. Penguin Group.
Halsman, P. (1973). Andy Warhol's factory [Photograph].
Livingstone, M. (1990). *Pop art: A continuing history*. Abrams Inc.
Musin, N. (Choreographer), & Odier, C. (Creative Director). (2007, March 15). *Trans-Warhol [Opera-dance]*. Bâtiment des Forces Motrices.
Otty, L. (2005). *The no man show: Technology and the extension of presence in the work of Andy Warhol*. eSharp, 5. University of Edinburgh.
Pallasmaa, J. (2005). *The eyes of the skin: Architecture and the senses*. Academy Press.
Press, E. O. P., Hickey, D., & Bluttal, S. (2006). *Andy Warhol "giant" size*. Phaidon Press.
Scherman, T., & Dalton, D. (2009). *Pop: The genius of Andy Warhol*. HarperCollins.
Schumacher, P. (2009). Parametricism: A new global style for architecture and urban design. *AD Architectural Design—Digital Cities, 79*(4).
Swenson, G. (1963). What is pop art? Interviews with eight painters, part 1. *Art News, 62*(7), 24–27, 58–59.
Szabó, M. (2018). Warhol and the Amiga: Exhibition review. *IEEE Annals of the History of Computing, 40*(4), 77–79. https://doi.org/10.1109/MAHC.2018.2877917
Warhol, A. (1962). Dance Diagram [2] (Fox Trot: "The Double Twinkle-Man") [Casein and graphite on linen].
Warhol, A. (1964). *Brillo boxes. [Silkscreen ink and house paint on plywood]*. Museum of Modern Art.
Warhol, A. (1977). *The philosophy of Andy Warhol: From A to B and back again*. Houghton Mifflin Harcourt.

Chapter 13
Transient Stages: Reimagining Performative Spaces Through Temporary Appropriation

Jose Antonio Lara-Hernandez and Irene Boles

Abstract This text focuses on the phenomena of temporary appropriation (TA) as a transformative agent within performative arts environments. The concept of TA, commonly associated with the versatile use of urban spaces by individuals or communities, here is recontextualized in the realm of performative arts. This chapter explores how TA catalyzes the evolution and cultural contextualization of performative spaces, drawing from historical precedents where temporary interventions have redefined the spatial and experiential parameters of performance. Highlighting this interdisciplinary dialogue, the chapter illustrates case studies where ephemeral installations, pop-up theatres, and impromptu stages have served as platforms for artistic expression, challenging traditional architectural constraints. These modalities of TA reflect a broader trend towards flexibility, spontaneity, and inclusivity in the design of art spaces. The discussion will extend to encompass the socio-cultural impact of TA on community engagement, highlighting how these temporary setups can democratize art consumption and foster new artistic communities. Additionally, the chapter will examine the potential of TA to anticipate future trajectories in the design of performative spaces, emphasizing sustainability and resilience. It will propose design principles informed by TA practices, which prioritize modularity, adaptability, and ecological responsiveness. Through this lens, TA emerges as a pivotal concept for rethinking performative spaces to be more reflective of diverse artistic needs and environmental considerations. In conclusion, this chapter advocates for a reconsideration of design methodologies to embrace the temporal and adaptive nature of artistic endeavors, positioning TA as a driver for innovation and a bridge between historical evolution and future opportunities in performative arts spaces.

J. A. Lara-Hernandez (✉)
Auckland University of Technology, Auckland, New Zealand
e-mail: antonio.lara-hernandez@aut.ac.nz

I. Boles
Ara Institute of Canterbury, Christchurch, New Zealand
e-mail: irene.boles@ara.ac.nz

A. Melis and M. Pelle (eds.), *Design for Performative Arts Spaces*,
https://doi.org/10.1007/978-3-031-98215-6_13

Keywords Urban spaces · Cultural contextualization · Historical precedents · Temporary interventions · Spatial redefinition · Ephemeral installations · Pop-up theatres · Impromptu stages · Artistic expression · Architectural constraints · Flexibility · Spontaneity · Inclusivity · Community engagement · Democratize art consumption · New artistic communities · Design principles · Modularity · Adaptability · Ecological responsiveness · Sustainability · Resilience · Innovation · Design methodologies

Introduction

In the last few years, the concept of temporary appropriation (TA) has regained importance because of its relevance towards topics such as identity (Lara-Hernandez et al., 2018), inclusion (Melis et al., 2020), social resilience and even mental health (Lara-Hernandez et al., 2019). Temporary appropriation is defined as "the temporary act in which people use public spaces to carry out individual or collective activities other than the purpose that the space was originally designed for" (Lara-Hernandez & Melis, 2018, p. 367). This definition is useful because it helps us to differentiate the temporary appropriation from the simple use of the public space. The former occurs when the individual (or group) feels identified with the environment, i.e. when the space is embedded with meanings. Following Relph's (1976) ideas, it is through these actions that the space becomes a place. According to Lara-Hernandez and Melis, (2018) there are three main categories in which TA could be grouped: (i) activities related to work or economy, (ii) leisure such as sports, games, and different forms of artistic expressions, and (iii) activities related to sacralisation or religious manifestations.

Performative arts put emphasis on actions, which is not simple because it is concerned with performativity. "The term performativity describes the interdependent relationship between certain words and actions – as when a word or sentence implies an action" (TATE, 2012).

According to Austin (1962), when someone says something, they are not just stating something (a locutionary act); they are doing something (an illocutionary act), and this action may have further effects (a perlocutionary act).

An illocutionary act refers to the performance of an act in saying something, in which it is not merely the utterance of words, but also includes the intention behind them. For example, saying "I connect" performs the action of connecting. In comparison, perlocutionary act refers to the effect or consequence that a speech act has on the feelings, thoughts, or actions of either the speaker or the listener. For instance, if the statement "I connect" makes someone feel connected, that's a perlocutionary effect.

In the context of performative art, this linguistic concept is borrowed to emphasise the action and the effects of the combination between human and art as a 'speech act' or 'utterance'. Thus, it suggests that performative art, when viewed through the lens of performativity, is an active, dynamic process where the performance is intertwined with its unfolding actions and its consequences, rather than just a representation or a static work.

Thus, the analysis of performative art could be similar to an agential cut as described by Barad (1996), in which they are seen as a learning process about phenomena, about specific material configurations of the world's becoming, thus, taking into account the researcher's role as an instrument of construction of the evidence, utilising a tripartite arrangement between objects, representations and knowers that produce knowledge (Lara-Hernandez, 2023).

Following this idea, performative art here is described as an "open work", which likely references Umberto Eco's (1989) idea of artworks that are incomplete and require the audience's participation to be fully realised. In this sense, performative art becomes a dialogue or a collaborative act between the artist, the audience, and the medium (in this case, possibly a computer), where the meaning and impact of the art extend beyond its initial presentation or "propositional content" (the basic content of the action).

The following paragraphs describe the performative spaces from a historical perspective.

Historical Context and Evolution of Performative Spaces

Since ancient times, the interplay of convergence of culture, social interaction, and architectural design have created immersive experiences through performative spaces Lara-Hernandez et al., 2018; Harvie, 2009). Such spaces have been subjected to TA, an act that dynamically influences both their spatial and experiential parameters. TA in the realm of performative arts is a manifestation of cultural, social, and environmental design flux (either natural or human-made), evidenced throughout history in various forms such as the Roman and Greek amphitheatres (Klar, 2006) the Aztec squares (Martínez, 2020), or the Shakespearean stages (Shurgot, 1998).

For instance, in ancient Greece, the amphitheatres were sites of TA where the polis—citizens of the city-states—would gather for entertainment and for collective participation in civic life purposes (Klar, 2006). In other words, the spatial dynamics of these amphitheatres were appropriated temporarily for different types of performances, including theatre, music, and political dialogue. These early performative spaces highlight the multi-dimensional aspects of TA, where the purpose of a space was reconceived temporarily to fit the needs of the performance and the audience. The design of such spaces, with open stages and tiered seating, facilitated a direct interaction between the performance and the spectators, a precursor to the modern reimagining of performative spaces.

Moving into the Mayan era, the Sacrifice's Cenote provides an illustrative example of TA influencing performative spaces. It is located in Chichen Itza, which was one of the most prosperous cities for the Mayan civilization that thrived from the third to twelfth centuries (Britannica, 2024). One notable feature of this ancient site is the Sacred Cenote, which held significant ceremonial importance during the Classic and Post-classic periods. Positioned 300 m north of the Mayan Castle, the central structure of Chichen Itza, the Sacrifice's Cenote has a diameter of 60 m, a water level of 22 m, and a depth of 13.50 m. This distinct geographical characteristic in the Yucatecan landscape, characterised by flat terrain and sparse vegetation, played a vital role in the Mayans' social and religious rituals, connecting their realm with the underworld (Cobos, 2023; Keller, 2006; Norman, 1842). The cenote's structure responded to the needs and customs of the time. Ceremonies and rituals were performed during the day, using natural light, and the audience surrounded the stage, creating an intimate experience that defied the rigid class structures of such period. This engagement between sorcerers, nobility, warriors, and spectators was a form of TA, transforming the social expectations of interaction and the utilization of space based on the performance.

Later in Europe, the Baroque period carried this forward, with opera houses and theatres exemplifying the fluidity of performative spaces. Scenography became an essential element of performance, with moveable scenery and machinery allowing for rapid changes in the setting. This mutability was a physical representation of TA, where the stage and its elements were in constant flux, morphing to support the narrative and emotional landscape of the performance (Kubalcik, 2004). It was flexible and versatile; therefore, the art of the performance dictated the form and function of space, rather than the other way around.

During the twentieth century, more pronounced instances of TA in performative spaces were produced, following the advent of avant-garde movements. Artist and performers challenged traditional norms and forms, utilising spaces in unorthodox manners (Sell, 2008). For instance, the Dadaists held performances in unusual venues like cabarets, and the Surrealists staged events in public spaces (Hopkins, 2004). These acts of TA were radical in that they took art out of conventional settings and into the public realm, where the distinction between performer and audience blurred, and the experience of the performance was as much about the reaction it elicited as it was about the action on display.

Recently TA has become integral to performance art, with artists using public spaces as a canvas for their work. Marina Abramović (Chalmers, 2008) and groups like the Guerrilla Girls (Withers, 1988) are a clear example in which pop-up performances in urban settings further demonstrate the multi-dimensional nature of TA. These ephemeral events create a sense of community and shared experience in the most mundane of places, transforming them into stages for artistic expression (Lara-Hernandez, 2023).

Temporary Appropriation in performative arts can be seen as a historical constant, shaping and reshaping the spaces where art is enacted and experienced. It has been a powerful agent in the evolution of performative spaces, dictating that such spaces are not static but are instead subject to the needs, messages, and experiences they aim to convey.

TA stands as a testament to the human capacity for adaptation, creativity, and interaction with our surroundings (Melis et al., 2024, Chap. 8). It invites us to consider performative spaces not merely as venues but as living, breathing extensions of the performances they host and the societies they serve. The following section illustrates contemporary case studies, describing the intrinsic relationship between arts and people through TA of public spaces.

Case Studies of TA in Performative Environments

Site-specific performances, a modern iteration of TA, describe performances that are crafted to fit particular locations, often outdoors, and redefine the spatial parameters by integrating the environment into the performance itself. These performances are not just held in these spaces but are created for them, resulting in a profound interaction between the space, the performance, and the audience (Andrews, 2022).

There are endless examples of TA in performative environments, but the relevance of such phenomena is commonly overlooked. Through TA, activities occurring in public spaces contribute to how urban identity is constructed (Bolio Arceo, 2012; Lara-Hernandez et al., 2018; Rapoport, n.d.). The following lines describe three examples from three different cities with different legal frameworks towards the use of public space, weather conditions, and cultures. These case studies describe ephemeral installations, pop-up theatres, and impromptu stages that challenge traditional architectural constraints and foster artistic expression.

The first is in San Francisco de Campeche, Mexico. Every year, in September, the local population commemorate the anniversary of the Black Christ of San Roman. Such festivity dates back more than 300 years and is celebrated in one of the most emblematic and traditional *barrios* (neighbourhoods) of the city. The main attraction is a performance representing the shipwreck, from which a crucified black Christ statue "survived" and arrived in the coast of the city. Later, a memorable catholic mass is held followed by a verbena.[1] Figure 13.1 illustrates how the street is transformed into an open theatre, where religious performances, music, dance, and socialisation take place.

Regarding the regulations over the use of the public space, in San Francisco de Campeche is not clear what activities are allowed or forbidden (Lara-Hernandez & Garcia-Villajuana, 2024).

Therefore, such regulatory framework in San Francisco de Campeche is similar to a case of legal pluralism, which advocates for plurality of legal systems. An scenario in which there is a co-existence of the state law, religious law, indigenous law, customary laws and local conventions (Lara-Hernandez et al., 2020; Marx & Kelling, 2018). In such a plural realm the law of the state is not necessarily the dominant one. Moreover, the state might not be able to have the capacity to enforce the law (McAuslan, 2006).

In terms of urban design, benches, portable chairs, and tables are strategically positioned, while the utilization of umbrellas as a cover is a contemporary design

[1] Popular festival with dance that is celebrated at night, outdoors and normally, because some festivity.

Fig. 13.1 Cristo Negro's Anniversary in San Francisco de Campeche, Mexico. Photo taken in July 2020. *Source* J. Antonio Lara-Hernandez

implementation to enclose the space, making the street a defined, intimate human-scale place.

The second example is the case for many streets in Venice, where the design of the built environment seems like a set, ready to be used. Figure 13.2 illustrates actors and players performing football acrobatic tricks during a video shot. The *sotoportego*[2] works as a 'portal' framing the player coming out triumphant with the ball, while the production team awaits on the other side. Thus, an example of TA in an impromptu stage.

Regarding the use of public spaces, Italian cities have a long tradition considering them as an scenario of social integration, allowing the right of association and the sense of a shared space. Including its appropriation as a collective space, the space for everybody (Romano, 2019). To our best knowledge this condition has only been compromised during times of crises such as war (Del Negro, 2002), the plague of the sixteenth century (Palmer, 1979), and the recent COVID-19 lockdown (Lara-Hernandez, 2023).

Music is an essential element of many performative events. For instance, in our third and last example, Fig. 13.3 shows a man playing the flute in Queen St. in Auckland, New Zealand, which is one of the most emblematic streets in the city.

[2] The space under the arcade (Treccani, 2024). Translation by the author.

Fig. 13.2 Influencers making video for social media in Venice, Italy. Photo taken in March 2024. *Source* J. Antonio Lara-Hernandez

Although that junction between Queen St. and Victoria St. is one of the most transited, the design configuration of the space (i.e. the column, the hall behind, the height to the ceiling, the materials, etc.) where the man is located amplifies the sound. Thus, by adding a chair the artist creates an ideal street stage, which is temporarily appropriated.

Culture is something often disregarded when it comes to how the public space is utilised. The following section briefly examines how TA affects community engagement.

Socio-cultural Impact of TA

The socio-cultural impact of TA is profound and multi-layered, influencing the fabric of urban life and the everyday experiences of citizens. TA, as explained elsewhere (see Lara-Hernandez & Melis, 2018; Lara-Hernandez, 2019, 2022) is a dynamic interaction between individuals and their surroundings, fundamentally tied to the creation of place and identity within the urban landscape.

Fig. 13.3 Man playing the flute in Queen St. Auckland, New Zealand. Photo taken in 2018. *Source* J. Antonio Lara-Hernandez

Historically, TA has manifested in the urban environment in a myriad of ways, serving as a crucial element for people's sense of belonging and connection to places. In practice, TA transpires through activities within the public spaces that form the stage for urban life. Moreover, these spaces become sites of social interaction and cohesion, creating a tapestry of relationships that is integral to the urban experience.

As explained by Lara-Hernandez (2019), the realm of leisure, encompassing arts and sports, is another category of TA. Here, public spaces become theatres for expression, canvases and stages for artists, and fields and playgrounds for athletes. The fluidity and spontaneity of such activities disrupt conventional norms and invite a collective reimagining of the cityscape. The arts, in particular, embody the ephemeral nature of TA, creating temporary but impactful connections among people and between people and places.

Sacralisation within TA represents a deep cultural element (Melis et al., 2024), adding spirituality to the urban tapestry. Public spaces become sanctified and defined through rituals and religious expressions, grounding communities in tradition while simultaneously allowing for the evolution of cultural practices in public view.

Through these lenses, TA is a vehicle for cultural expression, social engagement, and the construction of meaning in the urban environment. It is an act of resistance against the homogenization of public space, asserting the right to the city and the right to shape the city in ways that reflect its diverse inhabitants (Lefebvre, 1968). The transitory nature of TA notes the adaptability of urban spaces to meet human needs and desires, suggesting a responsive rather than prescriptive approach to urban planning and design (Melis et al., 2024).

Essentially TA is about empowerment. It challenges the status quo, allowing individuals and communities to claim agency over their environments (Lara-Hernandez & Chin, 2021). This action imbues spaces with cultural significance, rendering them as 'places' rich with social and historical connotations. Through TA, public spaces become a commons, a shared resource that reflects the collective memory and identity of its users.

TA as a Forecasting Tool for Future Design Trajectories

The analysis of TA is an essential tool to achieve more sustainable, inclusive, and versatile spaces that contribute to the formation of urban identity. Contrary to contemporary approaches based on principles that lead to over design urban spaces, the understanding of TA in a given urban context could lead to better architecture and urban design practice. For instance, studies in Mexico City Centre by Lara-Hernandez, Melis and Lehmann (2018), demonstrated that the number of TA activities observed was higher in "less designed" street sections compared to more stylish and recently renovated areas. Meaning that people's activities on the street occurred the most in areas where the design was less specific and mono-functional. Furthermore, in terms of urban planning, the authors noted that the diversity of TA is higher in areas of the street that are more distant from a square (or any open area). This provides valuable insights for inclusive and sustainable urban development because practitioners and developers could focus more (or less) in those areas that can foster TA, enhancing the relationship between people and place.

Synthesis of TA Principles for Performative Space Design

This work builds on the growing body of research focused on TA (Lara-Hernandez, 2019; Lara-Hernandez & Melis, 2018; Melis et al., 2024). Besides highlighting the role that culture plays on TA, an initial objective of the chapter was to try to describe design principles that could foster the emergence of TA in regards of performative arts. Taken from the literature (see Fig. 13.4) we could summarise our findings into the following design principles that enhance TA (Table 13.1).

After analysing the cases presented here under the scope of Principles for Performative Space Design we can highlight the following remarks:

The Cristo Negro Celebration takes place in the central square of Barrio de San Roman which is part of the heritage city centre of San Francisco de Campeche. Thus, it is designed at human scale, allowing flexibility while amplifying the human experience through the addition of decorative elements that spatially define the space such as the umbrellas and sound devices.

Venice is a peculiar case because on the one hand is up to the human scale (Choay, 2001; Lara-Hernandez, 2023), while on the other has a fractal urban fabric espouses

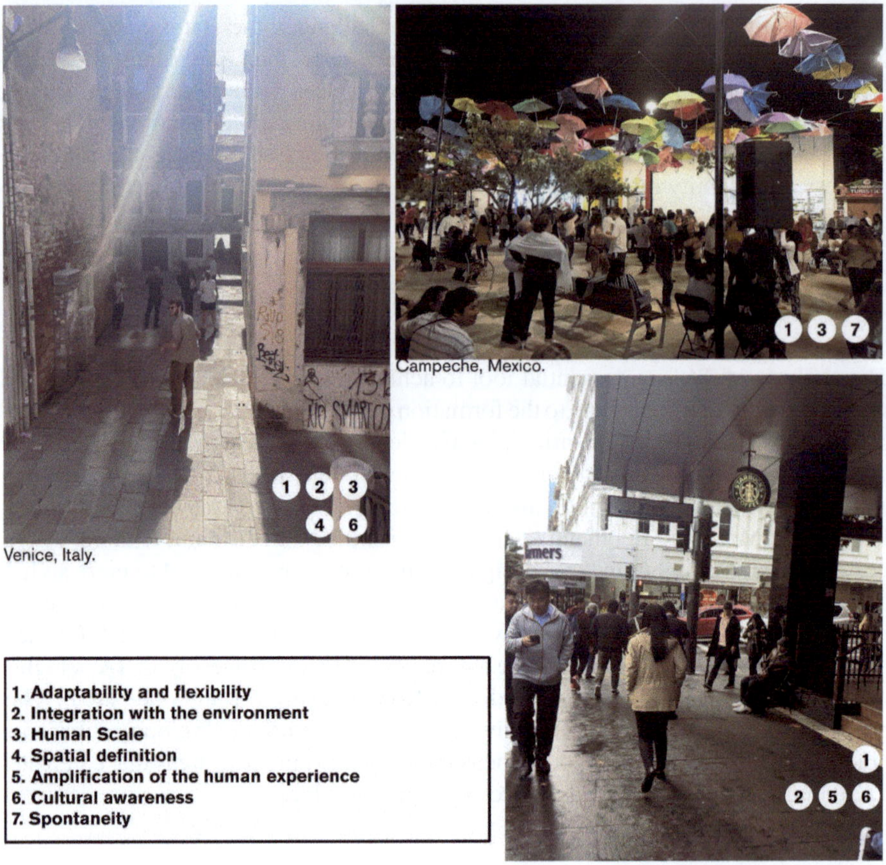

Fig. 13.4 TA principles assessment of the selected case studies. *Source* Authors

organic forms with rich detail at multiple scales either in its water channels and urban fabric (Boeing, 2018). It is worth to mention that our nervous system have evolved to expect fractal environments (Yang & Purves, 2003), because they have spatial variety, allowing their users have to develop sense of identity and community. This fractal design makes the city an ideal scenario for the user. Additionally, the absence of cars allows people to temporary appropriate public spaces freely.

Table 13.1 TA principles for performative space design

Versatility	Public spaces need to be adaptable to accommodate a variety of activities and events. The design should allow for temporary alterations or additions to facilitate different uses, such as performances or gatherings
Integration	Integration with the environment: public spaces should seamlessly integrate with their surroundings, including natural elements, as well as buildings. Performances crafted for specific locations should utilize and interact with the environment, blurring the lines between the performance space and the surrounding area
Human scale	Design elements should contribute to creating human-scale spaces that feel intimate and inviting. Benches, tables, low walls and other street furniture should be strategically positioned to encourage social interaction and engagement with the performance
Spatial delimitation	Spatial delimitation through design elements: walls, overhangs, canopies, and colonnades can help define and enclose spaces, transforming streets and public spaces into possible performance venues
Amplification	Design features should amplify the sensory experience of performances. Considerations like acoustics, lighting, and spatial configuration when designing public space, can enhance the audience's immersion in the performance
Cultural awareness	Design principles should take into account the cultural context and local traditions of the area where TA occurs. Understanding the significance of events such as religious festivals and traditional celebrations, can inform the design of spaces to accommodate such activities more effectively
Spontaneity	Public spaces should allow opportunities for spontaneity and improvisation, enabling impromptu performances or gatherings to take place. Urban design features should support creative expression through TA in all its forms

Source Authors

Conclusion and Advocacy for Temporal and Adaptive Methodologies

This chapter has explored the concept of Temporary Appropriation (TA) as a transformative force regarding performative arts. By examining historical precedents, contemporary case studies, and socio-cultural impacts, we have highlighted the dynamic relationship between TA and performative spaces. TA emerges not only as an indicator for analysing performative art in physical spaces but also as a catalyst for social interaction, cultural expression, and community engagement.

Historically, TA has been pivotal in shaping performative spaces, from ancient amphitheatres and Mayan cenotes to modern pop-up theatres and urban performances. These examples illustrate how spaces have been temporarily adapted to meet the evolving needs of performances and their audiences, fostering a more varied and immersive experience. Contemporary instances of TA, such as street performances and site-specific art, continue this tradition, challenging conventional architectural constraints and promoting a more inclusive approach to the use of public spaces.

The socio-cultural impact of TA underscores its role in democratising art consumption and fostering new artistic communities. By enabling spontaneous and flexible use of the public space, TA supports a more inclusive and responsive urban landscape. This versatility is crucial in addressing the diverse and ever-changing needs of urban populations, enhancing the resilience and sustainability of performative spaces.

Looking towards the future, TA offers valuable insights for the design of performative spaces. Design principles informed by TA emphasize modularity, adaptability, and ecological responsiveness. These principles advocate for spaces that are not only functional but also capable of evolving with their users, reflecting the temporal and adaptive nature of artistic endeavours.

Overall, the observation of TA could inform future design trajectories emphasizing sustainability and resilience by considering the social landscape of the place before undertaking any urban intervention. The study of TA prior the implementation of participatory design practices could make more inclusive and informed urban interventions. Such characteristics in urban interventions usually leads to more lively and authentic public spaces.

In conclusion, this chapter advocates for a rethinking of design methodologies to embrace the temporal and adaptive nature of performative arts. By positioning TA as a driver for innovation, we can bridge historical context and future opportunities, creating performative spaces that are more reflective of diverse artistic needs and environmental considerations. TA encourages us to view performative spaces not as static venues but as dynamic, living extensions of the performances they host and the communities they serve.

References

Andrews, A. J. G. (2022). *Know your place: An investigation of site-specific performance as an exchange of hospitality between artist and audience* (Doctoral thesis, University of Leeds). White Rose eTheses Online. https://etheses.whiterose.ac.uk/32896/

Austin, J. L. (1962). *Speech acts.* Oxford University Press.

Barad, K. (1996). Meeting the universe halfway: Realism and social constructivism without contradiction. In *Feminism, science, and the philosophy of science* (pp. 161–194). https://doi.org/10.1007/978-94-009-1742-2_9

Boeing, G. (2018). Measuring the complexity of urban form and design. *Urban Design International, 23*(4), 281–292. https://doi.org/10.1057/s41289-018-0072-1

Bolio Arceo, E. (2012). *Urban transformations and place-identity: The case of Merida.* Oxford Brookes University.

Chalmers, G. (2008). Marina Abramović and the redefinition of performance art: Boundaries, presence, and public space. *Journal of Contemporary Art Theory, 15*(2), 119–136.

Choay, F. (2001). Historic heritage and the contemporary culture industry. In *The invention of the historic monument* (pp. 138–163). Cambridge University Press.

Cobos, R. (2023). Chichen Itza: Sacred Geography and Ritual Use of the Sacred Cenote. *Journal of Mesoamerican Studies, 31*(2), 211–238. https://doi.org/10.32473/jms.v31i2.2023.001

Del Negro, P. (2002). *Il 1848 e dopo Storia di Venezia.* Treccani Istituto. https://www.treccani.it/enciclopedia/il-1848-e-dopo_(Storia-di-Venezia)/

Eco, U. (1989). *The open work* (A. Cancogni, Trans.; D. Robey, Intro.). Harvard University Press. http://nedimago.com/iconology/resources/books/Eco_Umberto_The_Open_Work.pdf

Encyclopedia Britannica. (2024). *Chichen Itza.* In Britannica.com. https://www.britannica.com/place/Chichen-Itza

Harvie, J. (2009). *Theatre & the city.* Palgrave Macmillan. https://doi.org/10.1007/978-1-137-04925-5

Hopkins, D. (2004). *Dada and Surrealism: A Very Short Introduction.* Oxford University Press.

Klar, L. (2006). Origins of the Roman scaenae frons and the architecture of triumphal games in Republican Rome. In Welch, K. (Ed.), *The Roman Amphitheatre: From Its Origins to the Colosseum* (pp. 135–163). Cambridge University Press.

Keller, A. (2006). Sacred Cenote at Chichén Itzá: Ritual Practices and Archaeological Evidence. *Journal of Mesoamerican Archaeology, 18*(3), 233–258.

Kubalcik, S. (2004). *Renaissance and Baroque stage technology and its meaning for today's theatre* (Master's thesis, Edith Cowan University). Edith Cowan University Research Online. https://ro.ecu.edu.au/theses/792/

Lara-Hernandez, J. A., Melis, A., & Lehmann, S. (2018). *Temporary appropriation and social sustainability: The case of the historic centre of Mexico City.* APRU Sustainable Cities and Landscapes 2018. https://apru.org/

Lara-Hernandez, J. A. (2019). *Temporary appropriation: Theory and Practice of the Street* (Issue August 2019) [University of Portsmouth]. https://researchportal.port.ac.uk/portal/en/theses/temporary-appropriation(62a30252-cbb1-457a-b751-52b66176d8d7).html

Lara-Hernandez, J. A. (2022). Resilience, architectural exaptation, and temporary appropriation. In M. Carta, M. R. Perbellini, & J. A. Lara-Hernandez (Eds.), *Resilient communities and the Peccioli Charter: Towards the possibility of an Italian Charter for resilient communities* (pp. 257–264). Springer International Publishing. https://doi.org/10.1007/978-3-030-85847-6_27

Lara-Hernandez, J. A. (2023). Emergent temporary appropriation versus lockdown regulations: The case of Venice. In *Sustainable Cities and Society, Recent Trends on Sustainable Smart Cities: Proceedings of IEEE ISC2 2022.* https://doi.org/10.1016/j.scs.2023.104854

Lara-Hernandez, J. A., & Chin, J. (2021). Sense of community and spatial agency: Key elements of resilient communities. In M. Carta, M. Perbellini, & J. A. Lara-Hernandez (Eds.), *Resilient communities and the Peccioli Charter towards the possibility of an Italian Charter for resilient communities Volume di Approfondimento del Catalogo del Padiglione Italia alla XVII Biennale di Architettura di Venezia.* Springer Nature.

Lara-Hernandez, J. A., & Garcia-Villajuana, A. (2024). *Emergent heritage: Co-creative NGO's work and selective protective government.* FARO.

Lara-Hernandez, J. A., & Melis, A. (2018). Understanding the temporary appropriation in relationship to social sustainability. *Sustainable Cities and Society, 39C*, 1–14. https://doi.org/10.1016/j.scs.2018.03.004

Lara-Hernandez, J. A, Melis, A., & Coulter, C. M. (2020). Temporary appropriation and informality: A nuanced relationship. In J. A. Lara-hernandez, A. Melis, & J. Thompson (Eds.), *Temporary appropriation in cities: Human spatialisation in public spaces and community resilience.* Springer Nature Limited.

Lara-Hernandez, J. A., Melis, A., & Lehmann, S. (2019). Temporary appropriation of public space as an emergence assemblage for the future urban landscape: The case of Mexico City. *Future Cities and Environment, 5*(1), 1–22. https://doi.org/10.5334/fce.53

Marx, C., & Kelling, E. (2018). Knowing urban informalities. *Urban Studies, 1*(16). https://doi.org/10.1177/0042098018770848

Martínez, M. (2020). Ephemeral architecture and public space: Cultural heritage in the squares of the Aztec civilization. *Journal of Architectural History and Theory, 14*(2), 101–118.

McAuslan, P. (2006). Legal pluralism as a policy option: Is it desirable, is it doable? In *Land Rights for African Development From Knowledge to Action, 2005* (pp. 9–10). http://www.capri.cgiar.org/wp/..%5Cpdf%5Cbrief_land-04.pdf

Melis, A., Lara-Hernandez, J. A., & Thompson, J. (2020). *Temporary appropriation in cities: Human spatialisation in public spaces and community resilience* (1st ed.). Springer Nature Limited. https://doi.org/10.1007/978-3-030-32120-8

Melis, A., Pievani, T., & Lara-Hernandez, J. A. (2024). *Architectural exaptation: When function follows form.* Routledge.

Norman, B. M. (1843). *Rambles in Yucatan: Or, Notes of Travel Through the Peninsula, Including a Visit to the Remarkable Ruins of Chi-Chen, Kabah, Zayi, and Uxmal* (2nd ed.). New York: J. & H.G. Langley.

Palmer, R. J. (1979). L'azione della Repubblica di Venezia nel controllo della peste. Lo sviluppo della politica governativa. In *Venezia e la peste, 1348–1797* (pp. 103–110). http://opac.regesta-imperii.de/id/264266

Rapoport, A. (n.d.). Australian aborigines and the definition of place. *Nature and Human Nature, 4*, 1–14.

Relph, E. (1976). Place and placelessness. In *London: Pion* (Vol. 1, 2nd ed.). Pion Limited.

Romano, M. (2019). *La città delle donne: Desiderio e bellezza.* La nave di Teseo.

Sell, M. (2008). *Avant-Garde Performance and the Limits of Criticism: Approaching the Living Theatre, Happenings, and the Avant-Garde.* University of Michigan Press.

Shurgot, M. W., & Owens, M. E. (1998). *Stages of Play: Shakespeare's Theatrical Energies in Elizabethan Performance.* New York: AMS Press.

TATE. (2012). *Performativity.* TATE. https://www.tate.org.uk/art/art-terms/p/performativity#:~:text=

Treccani. (2024). *Sotoportego.* Treccani Instituto. https://www.treccani.it/vocabolario/ricerca/sotoportego/

Withers, J. (1988). Feminist interventions in the art world: The Guerrilla Girls' use of parody and protest. *Art Journal, 47*(2), 33–38. https://doi.org/10.2307/776964

Yang, Z., & Purves, D. (2003). A statistical explanation of visual space. *Nature Neuroscience, 6*(6), 632–640. https://doi.org/10.1038/nn1059

Chapter 14
Embracing Technological Disruption and the Fourth Dimension in Art and Performative Art Spaces

Nicolas Turchi

Abstract This chapter investigates the intersection of art, architecture, and technology, emphasizing time as a transformative fourth dimension in creative spaces. This chapter highlights how advancements in fields like AI, VR, AR, and biotechnology are reshaping the role of art and architectural spaces, fostering dynamic interactions between creators, participants, and their environment. Turchi explores the synergies between art and architecture, considering installations as living systems and the influence of bioart on spatial design. He underscores how AI technologies, especially GANs, can create adaptive and responsive experiences, positioning audiences as active co-creators within augmented environments. Further, as real-time digital interactivity expands, the "active observer" model in exhibitions and performative spaces emerges, blurring boundaries between spectator and participant. Blockchain and NFTs add a temporal dimension, revolutionizing digital art through verifiable ownership and customizable access. Through case studies, including nt-ar's *Museum of Emotions* (MOE), Turchi advocates for a multidimensional approach to spatial design, merging temporal and spatial principles for immersive, participatory experiences that redefine architectural roles and cultural engagement in the digital era.

Keywords Technological disruption · Fourth dimension · Art and architecture · Technology integration · Time as transformative · AI (artificial intelligence) · VR (virtual reality) · AR (augmented reality) · Biotechnology · Creative spaces · Dynamic interactions · Living systems · Bioart · Spatial design · Adaptive experiences · Responsive environments · Active co-creators · Real-time digital interactivity · Active observer model · Spectator-participant boundaries · Blockchain · NFTs (non-fungible tokens) · Verifiable ownership · Customizable access · Case studies · Museum of Emotions (MOE) · Multidimensional

---

N. Turchi (✉)
Director at nt-ar, Dubai, UAE
e-mail: n.turchi@nt-ar.com; nturchi@aud.edu

Assistant Professor at American University in Dubai, Dubai, UAE

Adjunct Professor at Univeristy of Bologna, Bologna, Italy

A. Melis and M. Pelle (eds.), *Design for Performative Arts Spaces*,
https://doi.org/10.1007/978-3-031-98215-6_14

approach · Temporal and spatial principles · Immersive experiences · Cultural engagement · Digital era · Performative art spaces · Innovative installations · Creative interaction · Redefining architectural roles

Art and Architecture, Time as a Discriminator

In the ever-evolving landscape of art, time stands as the elusive fourth dimension, offering artists a dynamic canvas upon which to weave narratives, evoke emotions, and transcend traditional boundaries. As we begin exploring the longer-term rear effects of the second digital turn (Carpo, 2017), the significance of time in art gains new dimensions, shaping not only how we perceive and create but also how we interact with and experience artistic expressions. Arguably, a significant point of collision is the clash between disciplines that follow different timelines. While art grapples with an increasingly ephemeral and open-ended temporality, receptive to external injections of meaning and interpretation, its container, architecture, remains tethered to a semblance of a-temporality. Defined by the superstructures underlying its operational foundations, architecture often finds itself cast into the role of a sterile backdrop, divorced from the fluidity and dynamism inherent in contemporary artistic expressions. Conversely, when a large-scale object is designed as a temporary structure, it conventionally escapes any architectural definition, being labeled as an 'installation'.

Amid this tension, opportunities for dialogue and collaboration between art and architecture abound, especially with the advancement of new technologies. As architects embrace digital tools and techniques to envision new forms of spatial experience, they are increasingly attracted to the temporal complexities inherent in digital art. Likewise, digital artists aim to infuse their creations with materiality and presence, inspired by the spatial dimensions of architecture. By navigating this complex interplay between art, architecture, and technology, it becomes evident that the convergence of these disciplines offers fertile ground for exploration and experimentation. Through collaborative projects, interdisciplinary research, and shared dialogues, artists and architects are breaking down traditional barriers and forging new paths in the exploration of temporal design.

Art and Technology—Bioart

Among the emerging technologies contributing to the disruption of the art exhibition and performance space, it is worth mentioning Virtual and Augmented Reality (VR and AR), Artificial Intelligence (AI), Blockchain, Biotechnologies, and Bioart. Technical advances in biotechnologies and a renewed consciousness of environmental sustainability have pushed artists to explore the intersection of science, art, and technology in novel ways. The 'BioHacking' movement (Royal Society of Biology, n.d.) engages scientists and artists experimenting with biology outside of traditional laboratory settings. If further explored and connected to the architectural capacities of the 'art container,' these endeavors can redefine the concept of buildings as living organisms, fostering a holistic entanglement with the art piece that could take on different relationships existing in nature. Symbiotic relationships (Mutualism, Commensalism, Parasitism), Predation relationships, Competition, Mutual Exclusion relationships, and Facilitation (Stachowicz, 2001) represent dialectical models that share unexpected analogies with the built environment. Facilitation is a peculiar relationship that deserves further investigation when it comes to the container-contained relationship in art and architecture. It involves one species altering the environment in a way that benefits another species. For example, certain plants may improve soil fertility, creating favorable conditions for other plant species to grow. Similarly, some species may provide shelter or protection for others, facilitating their survival (Bulleri et al., 2015). This type of relationship echoes on a larger scale when the life-cycle of a building is expanded by hosting particularly successful exhibitions and events. Additionally, art pieces may make use of the container in a mutual re-adaptation of the content, so that the narrative becomes inextricable from its stage. This relationship slightly varies from Mutualism; despite both being positive interactions, Facilitation can be asymmetrical and involves one species altering the environment in a way that benefits another species, without necessarily receiving direct benefits in return. Unlike mutualism, it does not necessarily involve a direct exchange of resources or services between the interacting species. EcoLogicStudio provides an interesting example of this concept within the realm of Art and Architecture. In their recent publication 'Biodesign in the Age of Artificial Intelligence: Deep Green,' the founders of the studio C. Pasquero and M. Poletto explore technology in art and architecture as an emergent force deeply intertwined with biological processes, invoking a profound reevaluation of the relationship between human and non-human agents in the creative act (Pasquero & Poletto, 2023).

Art and Technology—Artificial Intelligence

While exploring biotechnologies and bio art offers a fascinating glimpse into the potential symbiotic relationships between art and nature, it also prompts us to reconsider the temporal dimension within the art space. This temporal reimagining finds resonance in the realm of Artificial Intelligence (AI), where algorithms and machine learning models operate within a framework of continuous adaptation and evolution. In recent years, AI has emerged as a powerful tool for artists and architects, offering new avenues for creative exploration and expression. From generative algorithms that autonomously produce intricate artworks to immersive installations that respond in real-time to viewer interactions, AI has the potential to revolutionize the way we conceive, create, and experience art. Generative adversarial networks (GANs) represent a groundbreaking advancement in artificial intelligence, particularly within the realm of creative expression. These algorithms consist of two neural networks—a generator and a discriminator—engaged in a continuous feedback loop, akin to a dance of creation and critique. What makes GANs particularly fascinating from a temporal perspective is their ability to capture and replicate the temporal dynamics inherent in artistic mediums. For example, in the generation of visual art, GANs can learn not only to produce static images but also to simulate motion and temporal progression. By analyzing sequences of images or videos, GANs can learn to generate fluid animations that evolve over time, capturing the nuances of movement and change. The convergence of GANs logic with VR and AR technologies holds the promise of a transformative redefinition of the art and exhibition space, ushering in a new era where spectators are active participants in the creation of artistic experiences. No longer passive spectators, audiences become dynamic agents capable of shaping reality itself, echoing the paradigm of the 'active observer' in contemporary physics (Ben-Ya'acov, 2021). In this symbiotic relationship between technology and audience, GANs-powered augmented environments serve as the canvas for collaborative creation, where the boundaries between creator and spectator blur. As viewers engage with immersive digital artworks, their interactions leave indelible traces, influencing the evolving narrative and aesthetic landscape in real time. This fusion of artistic expression and audience agency not only enriches the spectator experience but also opens up unprecedented possibilities for co-creation and collective meaning-making within the art space (Turchi, 2021) (Fig. 14.1).

Fig. 14.1 Nicolas Turchi nt-ar augmented experiences mixed media

Art and Technology—AR and VR

As AI seamlessly blends with immersive technologies that augment our experience of space, we are entering a scenario that seems to have advanced along the lines of space distortion tricks described by Venturi as 'the architecture of communication over space' (referring to places whose spatial qualities can be described through the communication strategy of the space itself) and 'the interior oasis' as identified in the 'Las Vegas Lighting' chapter: "The gambling room is always very dark; the patio, always very bright. But both are enclosed: The former has no windows, and the latter is open only to the sky. The combination of darkness and enclosure of the gambling room and its subspaces makes for privacy, protection, concentration, and control. The intricate maze under the low ceiling never connects with outside light or outside space. This disorients the occupant in space and time. One loses track of where one is and when it is. Time is limitless, because the light of noon and midnight are exactly the same. Space is limitless, because the artificial light obscures rather than defines its boundaries. Light is not used to define space. Walls and ceilings do not serve as reflective surfaces for light but are made absorbent and dark. Space is enclosed but limitless, because its edges are dark…The lighting is antiarchitectural." (Venturi et al., 1972).

We can now witness the dissolution of the linear and unidirectional relationship between performer/performance and audience. The latter, entrusted with the role of the 'active observer,' unlocks a new level of engagement. While collective efforts to enhance immersion in the experience and consumption of art and performances have been widespread, they have often fallen short of ambitious goals. It is no longer sufficient for the active observer to be regarded as a mere spectator. Audiences increasingly demand active engagement while experiencing art, design, and performances. This is where more advanced real-time AI-aided AR/VR environments can play a pivotal role in setting a new standard for the exhibition space. The delicate balance lies in the relationship between fully virtual and partially augmented environments. While a completely digital environment may still encounter resistance from older generations, a blended mode of consumption appears to appeal to both younger generations and those with a more traditional view. Deloitte's 17th annual Digital Media Trends Survey revealed that in the U.S.: "Watching TV shows and movies at home is no longer the dominant, "go-to" activity it once was—especially with younger generations that are more evenly dividing their entertainment time across TV shows and movies, user-generated content (UGC) on social media services, and video games. They seek entertainment, connection, immersion and utility." (PR Newswire, 2021).

Art and Technology—Blockchain and NFTS

As the spatial dimension becomes increasingly blurred, both through the amalgamation of media and stages, and the ostentatious pursuit of immersion in renovated traditional spatial settings, the temporal dimension emerges as the new yardstick for many aspects of art consumption. This is evident in social media platforms, where the number of times the audience "stops" a reel contributes to determining engagement and subsequently influences how potential content will be further distributed. Similarly, the time spent in an exhibition and the extended engagement within a certain event, whether in the form of a virtual extension through a mailing list or the user data shared with the venue, are becoming critical metrics. Moreover, time as rhythm emerges as the new spatial dimension. While in the past, a lengthy walk to a particular art piece, a long staircase, or a crowded room might have deterred me from engaging with the artwork, now, a non-responsive webpage, an intrusive cookies request, or a lag in the AI's response to my attempt at active participation could have the same effect.

This shift towards the temporal dimension underscores the evolving nature of art consumption, where the flow and rhythm of engagement are as crucial as the physical space in shaping the viewer's experience. In this digital age, where attention spans are fleeting and expectations for immediate gratification are high, mastering the art of temporal engagement is essential for artists, designers, and performers alike. Just as architecture manipulates spatial dimensions to evoke specific emotions and reactions, the strategic management of time in digital environments can profoundly influence audience perception and interaction. This is exemplified by the contractual revolution brought about by blockchain and Non-fungible tokens (NFTs) in the digital art industry. Blockchain technology and NFTs provide a means for authenticating digital artworks and other creative assets over time. As the ledger is immutable and transparent, it creates a permanent record of ownership and provenance. While platforms might change (given the absence of a fixed site), the connection to time ensures that the authenticity and ownership of digital assets can be verified throughout their lifespan, contributing to their long-term value. In more detail, through the adoption of smart contracts, creators can embed dynamic functionalities into NFTs, such as time-limited access or unlockable content based on specific conditions. This temporal dimension adds depth and interactivity to the digital art experience, encouraging prolonged engagement and exploration. All the while, blockchain technology guarantees almost instantaneous direct peer-to-peer transactions and sales, transcending traditional barriers of time and geography, and fostering a more fluid market.

The Active Observer Role

Looking ahead to the near future, AI-powered artworks have the potential to create dynamic and interactive experiences for viewers. AI algorithms, capable of analyzing user input and environmental data in real-time, enable artworks to adapt and respond to audience interactions. By integrating blockchain and NFTs, these dynamic artworks can be tokenized and owned by individuals, empowering them to participate in and influence the evolution of the artwork over time.

Similarly, in a fast-paced and over stimulating environment, the 'active observer' will emerge as the protagonist, particularly in the realm of performative arts. Here, the fusion of augmented reality (AR), artificial intelligence (AI), and dynamic or responsive architectural design will challenge the traditional concept of theater spaces. Augmented reality technologies allow performers to break free from the constraints of traditional stage settings, enabling interaction with virtual objects and environments in real time. With the assistance of AI-driven generative algorithms, performers can dynamically adjust their movements, gestures, and expressions to synchronize with virtual elements, seamlessly integrating live action with digital augmentation. Furthermore, AR-powered audience interfaces empower spectators to engage with performances in unprecedented ways, reinforcing their role as active participants in the artistic experience. The recent release of GPT-4 promises real-time translations, annotations, and contextual information about our surrounding environment, potentially disrupting performative arts and enhancing comprehension and appreciation for diverse audiences. Additionally, AI algorithms can analyze audience reactions and preferences, dynamically adjusting the narrative flow, pacing, and emotional tone of performances to maximize engagement and impact. This symbiotic relationship between performers, audiences, and technology will enhance accessibility and deliver a customized experience. While this presents a positive development, it also poses a significant risk to the traditional social implications of performative spaces. "A society awash with images will experience a consequent reduction in social and political sensibilities, as the intoxication of the image leads to a lowering of critical awareness." (Leach, 1999). In contemporary times, the concept of sharing experiences has undergone a fundamental transformation. While many architects have endeavored to enrich spaces with elements such as intricate geometries, dynamic lighting, and vibrant color schemes, to draw people back to physical environments for communal experiences, the emergence of new technologies and social media platforms has introduced additional layers of filters between observers and reality. This has fundamentally altered the idea of collective experience, where each individual's viewpoint is mediated through their own unique lens, resulting in a constructed semblance of shared experiences. As a result, social dynamics are being redefined, and new roles for designed spaces are being delineated. Events such as the Eurovision Song Contest, which have witnessed a resurgence in recent years, epitomize this trend by meticulously planning venues to accommodate not only live audiences but also media perspectives. By leveraging digital enhancements and special effects, these events create a mesmerizing spectacle that deviates considerably from the unembellished experience of attending the live event.

To effectively tackle the social ramifications and significant disruptions brought about by these technologies, architects, designers, and an expanding array of professionals must collaborate within multidisciplinary teams. These teams will require diverse expertise to facilitate the seamless integration and connection of both physical and digital spaces. This shift necessitates moving away from traditional spatial considerations towards embracing new vocabulary and principles derived from the temporal dimension. By strategically incorporating temporal concepts, these teams will be empowered to create immersive experiences within static environments, thus reshaping the fundamental role associated with spatial design.

A Case Study—MOE by nt-ar

At nt-ar, as part of our ongoing design research, we continuously question the nature of spaces and their relationships with users, evaluating them against emerging trends and technologies. nt-ar's project submission for the Museum of Emotions (MOE) exemplifies a departure from passive observation, beckoning visitors to embark on an immersive existential odyssey while setting a new relationship with exhibited art pieces. Within the confines of the ring-shaped museum, a symbiotic interplay unfolds between the curated galleries and the surrounding natural environment, fostering an ever-evolving experience. The main idea behind MOE was to defy conventional cultural consumption by empowering visitors to engage deeply with art on a profoundly personal level. The convergence of negative and positive emotions guides individuals on an introspective voyage of self-discovery. Through this emotive exploration, visitors establish a profound connection with the space, transitioning from passive observers to integral participants in the museum's narrative. At MOE, the boundaries between art and life dissolve, enabling visitors to redefine their relationship not only with the space but also with themselves. Each visit promises a singular and unforgettable journey, wherein the museum transcends its role as a mere venue—it becomes a mirror reflecting the complexities of the human experience, a sanctuary for the soul, and an immersive encounter leaving an indelible mark on all who traverse its hallowed halls (MOE, nt-ar, n.d.). This project, among many others, defines a new trend in research-driven and exploratory practices aimed at tackling the pressing questions discussed above. The incorporation of emerging technologies into spatial experiences, coupled with a heightened awareness of the temporal dimension and its implications, promises to chart new paths, expanding architectural vocabulary to align with the evolving discourse introduced by related disciplines.

References

Ben-Ya'acov, U. (2021). The architecture of communication over space. *Journal of Physics: Conference Series, 1956*, 012019.

Bulleri, F., Bruno, J., Silliman, B., & Stachowicz, J. (2015). Facilitation and the niche: Implications for coexistence, range shifts, and ecosystem functioning. *Functional Ecology*. Advance online publication.

Carpo, M. (2017). *The second digital turn: Design beyond intelligence*. The MIT Press.

Leach, N. (1999). *The anaesthetics of architecture*. The MIT Press.

MOE, nt-ar. (n.d.). Museum of Emotions project description.

Pasquero, C., & Poletto, M. (2023). *Biodesign in the age of artificial intelligence: Deep green*. Routledge

PR Newswire. (2021, March 9). The quest for connection: Younger generations look to user-generated content and video games to find value, meaning, and personal fulfilment. Retrieved from https://www.prnewswire.com/news-releases/the-quest-for-connection-younger-generations-look-to-user-generated-content-and-video-games-to-find-value-meaning-and-personal-fulfillment-301798079.html

Royal Society of Biology. (n.d.). *Focus on biohacking*. Retrieved from https://www.rsb.org.uk/biologist-features/focus-on-biohacking

Stachowicz, J. (2001). Mutualism, facilitation, and the structure of ecological communities. *BioScience, 51*(3), 235–245.

Turchi, N. (2021). The architecture of spacetime: Memory as a project. *diségno, 8*, 97–108.

Venturi, R., Brown, D. S., & Izenour, S. (1972). *Learning from Las Vegas: The forgotten symbolism of architectural form*. The MIT Press.

Chapter 15
A Stolen Wall from the Venice Biennale

Lanfranco Aceti

The two most powerful warriors are patience and time. Lev Nikolayevich Tolstoy

Abstract This chapter provides an introspective exploration of art's relationship with space, freedom, and institutional critique. It reflects on the artist's experience with the Italian Pavilion at the Venice Biennale from 2019 to 2021, during which he conceptualized "stealing" a wall. This creative act symbolized artistic appropriation and decentralization, as well as a rejection of conventional institutional constraints. Supported by curator Alessandro Melis, Aceti removed a temporary wall, which he re-contextualized as a performative object. Placing it in rural Italy, he used it as both a physical and symbolic space to engage with local sociopolitical histories, juxtaposing it against modern-day cultural dynamics.

Keywords Stolen wall · Venice Biennale · Lanfranco Aceti · Art and space · Freedom in art · Institutional critique · Italian pavilion · Artistic appropriation · Decentralization · Conventional constraints · Curator Alessandro Melis · Temporary wall · Performative object · Rural Italy · Sociopolitical histories · Cultural dynamics · Creative act · Re-contextualization · Physical space · Symbolic space · Juxtaposition · Modern cultural context · Art engagement · Artistic intervention · Community interaction · Critical reflection · Transformative practices · Institutional boundaries · Artistic expression · Social commentary

It is an opportune moment to reflect on the intricate set of circumstances that facilitated the development of an unconventional collection of artworks, alongside the adoption of unorthodox and performative curatorial and artistic practices. Throughout the triennium 2019–2021, time has been one of my most invaluable allies, enabling the completion of an extensive body of work that I had never before endeavored to undertake.

L. Aceti (✉)
Director, Museum of Contemporary Cuts (MoCC), Cassino, Italy
e-mail: lanfranco.aceti@gmail.com

A. Melis and M. Pelle (eds.), *Design for Performative Arts Spaces*,
https://doi.org/10.1007/978-3-031-98215-6_15

This journey commenced with the idea of metaphorically 'stealing' a wall from the Venice Biennale, allowing me to work freely from home without the need for constant physical presence. This approach also permitted me to forgo participation in the celebratory rites of the inauguration and the ongoing events of the Venice Biennale.

How a Wall Came to Be Stolen from the Venice Biennale

Occasionally, time permits the emergence of unorthodox practices through the gaps of interstitial artistic endeavors. These practices, though they may appear offbeat and out of place due to their deviation from the norm, offer opportunities to rethink established approaches. In numerous ways, I have resisted the urge to conform and become complicit with a conventional mode of art-making. The notion of bending art to institutional demands and creating curatorial compromises has never appealed to me.

From the outset of my career, I understood that certain compromises were inevitable. However, since I was engaged in the long-term pursuit of art history, I was instinctively certain that such compromises would neither benefit my artistic practice nor the artworks themselves.

Mindful of this, my participation in the 17th Architecture Biennale involved no compromises, but rather extensive discussions on the validity of artistic practices which challenged entrenched notions of the art object, the space, and identity (Frascina, 2011; Wang, 2020). This was made possible by the curator who supported my project, which others might have dismissed as unfeasible. The realization of these artworks was entirely due to the support of Alessandro Melis, the curator of the Italian Pavilion. His backing allowed the project to come to fruition, along with an ephemeral performative practice in the space of art creation—the studio—and the exhibition space—the museum. These practices were deeply rooted in my personal, uncommon, and almost esoteric research.

I was creating a body of artworks not solely as an aesthetic endeavor but also, in some cases, as documentation of inherited social and anthropological practices of living that span thousands of years.

The research for these artworks aimed to document what had been, what is, and what would never be again, creating an emotional engagement that continuously stretched between the personal and the universal. The artworks were produced within the architectural space of a vegetable garden in central Italy. This garden embodied the sedimentation of local anthropological farming cultures, juxtaposed against the imported, globalized transformation processes of post-postindustrial food production and farming.

The idea of 'stealing a wall' from the Italian Pavilion at the Venice Biennale became a subject of extensive discussions. Although a proper wall could not be removed due to the buildings being under the aegis and protection of the Italian Ministry of Culture and subject to strict legal protections, the conclusion of these

Fig. 15.1 Lanfranco Aceti, *A Stolen Wall from the Venice Biennale*, 2021. Cement bricks, mortar, and a visitor. Print on fine art Paper. Dimensions: 67 cm × 100 cm

lengthy conversations was that I could take away an internal wall. These internal walls are temporary and discarded every other year when deemed unusable and of no value.

Thus began the conceptual journey of the artwork *A Stolen Wall from the Venice Biennale* (see Fig. 15.1).

For me, it was an exciting endeavor that spoke to appropriation, decentralization, and freedom of expression. It also embodied an ephemeral performativity, which I chose to document solely as an artwork. The project remained peripheral; in hindsight, I suppose I could have generated more publicity around it. However, I was deeply focused on the process of art-making and the curatorial exchanges with Melis, leaving little room for considerations of spin and marketing. Having a wall stolen from the Venice Biennale was all that mattered to me and it brought me great contentment. I just needed to find a place for it.

Before placing the wall somewhere, it was important for me to understand what it meant, both artistically and curatorially, to declare that I had stolen a wall and that the curator had permitted me to do so. My initial thoughts centered on the necessity of opening up the process of stealing a wall to others, as for me, to steal also means to share. I have always believed in the idea of subtracting something from those who have an excess of it, but only if the redistribution of wealth would enable greater participation.

The pandemic's ravaging effects convinced me that the issue of collective participation could be deferred to a later stage. Thus, my personal engagement with the wall—both aesthetically, performatively, and curatorially—had to take precedence. This was essential if I truly wanted to realize and complete a project that required ten different stages and a rather large number of artworks.

Where Do You Place a Stolen Wall?

Curatorially, at least for the initial month or so, the relationship between the wall and myself was one of mutual observation and exploration. The wall had been situated in the midst of a countryside in Italy with little notoriety or claims to fame. Ravaged during World War II, it had been a focal point for socialists and fascists alike—approximately divided among the two ideologies, with few independent thinkers. The area had witnessed bombings and casualties by German, English, American, and French troops. Life in this small, somewhat bigoted countryside was juxtaposed with my personal memories and contemplations on the roles of rural and peripheral spaces, contrasted with the metropolitan environments of Rome and Naples.

It encapsulated performative collective labor and rituals governed by the natural cycles—a social nostalgia now overshadowed by new forms of asocial and individualistic consumerism.

It was a place where political discourse was colored by vivid imaginings of how politicians should be treated—gutted like fish, hanged like pigs, strung up like chickens, rabbits, or pigeons—each comment followed by suggestions of turning them into sausages, using them as manure for fields, or leaving them for insects and wild animals to devour. These remarks didn't come from the men, who were rather tame and somewhat juvenile in their political reasoning, but from the women. Their tongues wielded sharp barbs, each insult more outrageous than the last, often sparking uproarious laughter at the expense of one politician or another. It seemed like a competition for the most offensive and daring insult, marking a clear divide between 'us' and 'them'.

However, this performative space had been stifled by dwindling opportunities for social interactions and collective labor. The past and present were separated by sweeping changes, profoundly altered by the imperialistic globalized waves of Anglo-Saxon expansionism, the latest cultural manifestation of which is represented by the white cube (O'Doherty, 1999).

My political understanding of life emerged from those kitchens where different dishes were prepared at different times, and from those courtyards where various animals were slaughtered or different vegetables and grains were cleaned before consumption. It was a space of processes and performances that made life all the more fascinating.

Whenever an insult about a politician—local, national, or international—was blurted out during the many rituals of daily life that punctuated the year, hands would rise, and a large knife would often be brandished in the air. A collective performance ensued—a commedia dell'arte—where personal, familial, and local narratives intertwined. One infamous incident involved two women fighting a duel over a man with cold steel, each wielding a 'ptatora'—a farmer's tool used for cutting branches in central Italy, but also as a weapon. The consensus was always that they should have used their tools to chop up the man.

The reason I wanted to place the wall in contrada Campolargo in Piumarola, Villa Santa Lucia—now reduced to a non-place for all intents and purposes—was to establish a socio-political landmark of localized and practical matriarchalism. This stood in contrast to the globalized idealism of the 'chattering celebrity classes' whom, over time, I grew to despise as much as their political and intellectual counterparts.

My life has been a continual process of engagement and disengagement, akin to Giovanni Verga's constant attempt to fit into various molds, only to realize each time how little was shared and how little was in common. It has been a series of utilitarian relationships where, once a specific function or purpose that underpinned the 'friendship' ended, the relationship would instantly wither and die. I have left so many of these relationships in the global art world that I could fill the local cemetery, and then some.

As the pandemic took its toll—spanning 2019–2021 in Italy—the notion of having a deeply personal space in an area burdened by historical events and marked by environmental degradation due to political greed felt like a form of vindication. It was a testament to years of exploitation and destruction, capturing the local greed through documentation. Political critiques now aimed more sharply at corrupt local officials than at national figures, who, though still criticized, became less of a focal point. The abandonment of entire regions to industrial pollution for personal gain had become painfully evident, even to those who had hoped for change by voting for local councillors.

The mood turned dark, and criticisms grew more virulent towards familiar faces within the community. The shift was significant; it was no longer about distant and abstract political figures but about real individuals, flesh and blood, who had trampled upon acquaintances, friends, and family in their pursuit of personal advancement. Suddenly, the performative elements that once elicited laughter—international politicians are easily reduced to stereotypes—took on a darker and more somber tone when directed towards local individuals known by everyone.

The wall, situated in the heart of the land where artworks embodied remnants of ancient matriarchal values alongside the environmental and social devastation wrought by patriarchal capitalism, suddenly became a performative symbol of personal and local reckoning as the horror of the pandemic took hold.

The Architecture of a Stolen Wall and of the Performative Solace

The wall stolen from Venice was conceived as a white square space—a backdrop against which events could unfold, or not. It was designed and structured as a flat surface that could remain empty, serve as a backdrop, or be covered with artworks. It existed as a space of interstices and assemblages, within which an aesthetic narrative could evolve over time.

When one thinks of a wall, it's often envisioned as a flat element of a structure. For me, however, the wall was not merely a component of construction but a window through which the external environment could interact and influence the artistic production unfolding in front of and around it.

I had no intention of replicating the more or less successful attempts to bring the natural world into the white cube of galleries and museum spaces. This had already been done with myriad landscape paintings and installations that varied in their openness to the outside environment. The idea of sequestering nature and placing it within the sterile aesthetics of a white cube was not my aim. Such a process would anaesthetize and aestheticize nature, stripping it of its natural vitality and bending it to fit curatorial and institutional concerns primarily focused on entertaining and impressing the masses (O'Doherty, 2007).

For me, the wall had to serve a function beyond holding up images and acting as a backdrop. It was intended to allow the environment to influence how art itself was produced, acting as both a mirror and a constant reminder to the artist of the limitations of artistic creativity compared to the creativity of nature itself. Additionally, it served as a continual source of inspiration.

The wall had a simple architectural role: to protect the vegetable garden developing in front of it from strong winds, the prying eyes of curious audiences, and the scorching midday sun of July and August, providing shade for vegetables, animals, and humans alike.

The wall served to protect, defend, and conceal the life unfolding both in front of it and within its interstitial spaces, where lizards, ants, wasps, and various insects sought refuge. These worlds were not hostile but simply unseen by those who viewed the wall merely as a structure of bricks. Birds would swoop in and perch near the wall to snatch insects or probe these gaps in search of prey. Startled insects would take flight, often to be seized mid-air by the birds and carried away. At times, the wall provided a resting spot for birds, while other times it served as an observatory from which they could survey the vegetable patch below and indulge in a tasty morsel.

I wasn't concerned about the lack of audience engagement, a drawback for many artists—especially concerning buyers, critics, and collectors. Being unseen didn't negate the existence of the artworks; they existed without the obligation to engage an audience, particularly not in conventional terms. The landscape brimmed with life, just as my mind teemed with thoughts inspired by the unfolding of life, even amid death.

An interesting episode unfolded suddenly while we rested in the shade after lunch during the hottest hours of the day. A commotion caught our attention—a jay (*Garrulus glandarius*) had seized a small sparrow in its claws. Unexpectedly, all the other sparrows banded together as a flock and launched a counterattack against the predator, screeching and diving at it. Amidst the chaos, the jay released the small bird from its grip. We watched as the little sparrow fell to the ground, remaining motionless for a few moments before suddenly stirring and taking flight once more. These seemingly innocuous sparrows had united and saved one of their own.

It evoked memories of the countryside and its battles. These were tales of survival fights—gender, class, or political battles, all driven by the singular goal of surviving a harsh social environment in Italy. These stories, some unsuitable for children's ears yet never too distant, spoke of survival and cunning. They reminded me of the peasant uprisings against local landlords, spanning centuries and reaching back to Roman times. The harshness of life resonated with me through stories retold and performed, as well as through the literary depictions found in Sicilian and Sardinian works by authors such as Verga, Grazia Deledda, and Luigi Pirandello.

The sickle and pitchfork, innocently resting against the wall before or after toiling under the scorching sun in the fields, sometimes took on a more ominous meaning in my eyes. I recalled carrying water, wine, and beer to field laborers during the days of manual labor, while the kitchen bustled with lunch preparations and the glinting metal blades fell upon and harvested fields of wheat.

The vengeance of the people against abuse manifested in stories like the poisoning of a priest by his servant, who was both a sexual toy and a victim, or the local politician awaited at night to be shot from behind a wall in retribution for his misdeeds and thefts. Memories of such tales flooded in from an agricultural world far removed from the post-modern woke glamour of metropolitan cities like London, New York, Berlin, or Venice.

Young girls sent as servants at an early age to the homes of the wealthy often faced harassment and had to fend off advances from male household members. Those less fortunate, unable to defend themselves or lured by false promises of a better future, were destined to become pregnant and return to lives of poverty and hard labor, now burdened with raising a bastard child and adding another mouth to an already meager peasant's table.

A wall doesn't have to be just a wall; it doesn't need to be stripped of context and historical inheritance. The trend of decontextualization in the white cube gallery space has its limitations. By eliminating humanity, history, and everything else in favor of an imperial interpretation, made of a reduction of infinite realities, the white wall sharply focuses all of the viewer senses on something that might otherwise blend into its surroundings (Filipovic, 2014). Thus, it must be placed on a white wall to give it meaning—a sharp focus in a context of nothingness that creates something out of nothing. This process imbues art, whatever form it takes, with a sense of awe through selection, decontextualization, isolation, and display on a blank canvas. Nothingness is then filled with whatever one chooses to bring or is instructed to bring into the picture. The white cube is a corporate imperial whitewash.

But an object imbued with aesthetic nostalgia for its context loses significance if cultural inheritances are erased and unknown. Art and the world, to me, should be observed with curious eyes that slowly scan and absorb everything around them, much like blotting paper or the wide-open eyes of children. These eyes peer into the interstices, the cracks within walls, and discern other worlds and potential lives. Life becomes performative when it is actively engaged and has a space in which to unfold. This performance is facilitated by spaces that were once distant from the scrutiny of politicians and corporations but are now under the control of media that direct socially engaged mediated lives and dictate aesthetic values.

A wall is just a wall if we choose to see it as nothing more. Without our existential performances to infuse the displayed work with stories, emotions, life, and death—a picture remains just a picture, and a wall simply remains a wall, even if it was stolen from the Venice Biennale.

Art will always find another form to be displayed and performed elsewhere.

The Performance of Art on a Wall

The positioning of the artworks on the wall had its own complexities. The primary issue I faced was the relationship between the wall and the history of the institutions it represented. The need for artistic validation—as if the institutional charge of the wall would legitimize the artworks—was something I sought to break free from. It seemed that the multiplicity of perspectives and interstitial worlds should not be isolated but re-encapsulated in an understanding of 'being biennial' that had little or nothing to do with institutional critique by individuals entrenched within institutions. These individuals, whether motivated by affiliation, personal financial gains, or simply sycophantic tendencies, could not exercise a critique that would move in any direction other than that assigned by the institution itself.

"...[T]he role of the museum is precisely about performing art's function as publicity within a prescribed and always already compromised cultural space, in order to wrest from it a partial and contingent critical publicity, in terms of which a correspondingly mobile and perhaps strategic public might form," wrote O'Doherty (1999).

The lack of publicity, the absence of a compromised cultural space, and the absence of a public had the effect of making the wall almost an uncompromised locus. Any compromises present were those I brought to the wall myself. If art had to be framed within a particular period—the designated biennial—in order to exist, I had the power to extend that timeline to suit particular needs that had nothing to do with the institution. It seemed contrary to the unnatural order of the art world and its fixed times; thus, one way to break free was to avoid establishing a deadline. Imposing an end felt like imposing yet another artificial construct, which had little to do with art and much to do with extraneous reasons unrelated to the aesthetic trajectory I had embarked upon. The pandemic had been a blessing in this sense because deadlines were extended, stretched, and re-announced in a process that seemed aligned with nature itself rather than the artificial representation of natural processes and the performance of art gestures and narratives for collectors and publics with timelines imposed by external reasons not intrinsic to art itself.

This is the complexity of the space that opposes the white cube—a white cube that has become a space for the domestication and simplification of art rather than a locus where art subsumes within itself the complexity of our performative existence in a space and through a trajectory that we should redefine instead as the space for performing our time to live.

References

Filipovic, E. (2014). On the global white cube. *Oncurating, 22*, 63–84. Retrieved from https://www.on-curating.org/issue-22-43/the-global-white-cube.html

Frascina, F. (2011). White cube, white culture, white riot: Identity, ambivalence and non-belonging. *Third Text, 25*(5), 523–538.

O'Doherty, B. (1999). *Inside the white cube: The ideology of the gallery space*. University of California Press.

O'Doherty, B. (2007). *Studio and cube: On the relationship between where art is made and where art is displayed*. Princeton Architectural Press.

Wang, S. (2020). Museum coloniality: Displaying Asian art in the whitened context. *International Journal of Cultural Policy, 27*(6), 720–737.

Chapter 16
Architecture of Exhibitions: An Immersive Experience

Virginia Cucchi

Abstract History has taught us that, in its rigid, unilateral sectorization and categorization, it has often distorted representations, spreading homogeneous, misleading interpretations, marginalizing diversity, and stigmatizing a plurality of voices that would provide a more comprehensive vision. An evolution, mostly promoted by intellectuals and visionary artists, recognizing these limitations and the legitimate urge to reevaluate all cultural expressions, has advocated a change of perspective, stimulating more openness and flexible interaction. Societies need to become more inclusive: many institutions are revisiting and revising their own past, and exhibitions offer the possibility to present in a contemporary reading key, interwoven with the complex, multi-layered history, webs of echoes and references to new ways of non-traditional storytelling. The ephemeral character of the architecture of exhibitions reserves the privileged potential to experiment and set dynamic, engaging spatial stages. Architects, overcoming the rigorous canonic institutional orthodoxy and rethinking conventional language, have gradually re-contextualized and re-vitalized the exposition space from a simple depository of content to a relational space, through sensorial and intellectual stimulations. The exhibition '*device*', analyzed in all its poly-semantic nature, finds new allies, new contaminations animated by a different expressiveness that, broadening communication through multidisciplinary insights, allow to craft immersive journeys. New multi-sensorial environments, adopting diversified tools and an effective, accessible vocabulary, more suited to present day, arouse curiosity and affinities, actively connecting and establishing fruitful synergistic processes. The powerful convergence of different modalities, at the intersection of architecture, arts, science and technology, embracing cultural, political, ideological, social and economic issues, above aesthetics, brings with consistency visual clarity, reaching a diverse and global audience, triggering encounter and participation, and contributing to a dialectic exchange. The exhibition as a heterogeneous manifestation of an ideological and aesthetic concept, reformulated according to an ethos of horizontal transversality, acquires new vitality through authentic cooperation and integration, transforming itself into a workshop, a laboratory, a public space, where the spectator becomes a participant. As an example of this paradigm shift, the

V. Cucchi (✉)
Milan, Italy
e-mail: virginiacucchi.com@gmail.com

A. Melis and M. Pelle (eds.), *Design for Performative Arts Spaces*,
https://doi.org/10.1007/978-3-031-98215-6_16

17th and 18th editions of the Venice Architecture Biennale have been privileged for motivating and promoting, with continuity of intents, a widely hybrid contaminated experiential architecture.

Keywords Rigid sectorization · Cultural representation · Misleading interpretations · Marginalization of diversity · Plurality of voices · Reevaluation of cultural expressions · Change of perspective · Inclusive societies · Institutional revision · Contemporary readings · Multi-layered history · Non-traditional storytelling · Ephemeral architecture · Dynamic spatial stages · Overcoming institutional orthodoxy · Rethinking conventional language · Relational space · Sensorial stimulation · Intellectual stimulation · Poly-semantic nature · Multidisciplinary insights · Immersive journeys · Multi-sensorial environments · Accessible vocabulary · Curiosity and affinities · Synergistic processes · Convergence of modalities · Intersection of arts and science · Cultural issues · Dialectic exchange · Heterogeneous manifestation · Ideological concepts · Horizontal transversality · Authentic cooperation · Public space · Spectator participation · Venice Architecture Biennale · Hybrid experiential architecture · Paradigm shift · Engaging exhibitions

Venice Architecture Biennale 2021

The 2021 and 2023 Venice Architecture Biennale represent a significant, emblematic attempt to abandon rigidity of disciplinary boundaries and self-referentialism for a more collective dimension. Hashim Sarkis, curator of the 17th Exhibition, addressing the urgent, nebulous situation of global emergency, with the question 'How will we live together?' (Sarkis, 2021) has encouraged architects, with an implicit appeal to a more empathetic understanding of the discipline, to overcome dominant singular and incomplete views and face the demanding challenges, encompassing the broadest representation of different voices of the community, from experts to ordinary citizens, in order to find through everyone's expertise potential alternatives to the weaknesses of past models in the rapidly changing political, social and climatic scenario.

The perspective of an increasingly complex future has conferred a loud voice to an aphonic multilateralism, and the narration has started to become a plural story told with languages ranging from photography to installation, from sculpture to drawing, and from sound to performance.

The Italian Pavilion, in particular, has accepted the invitation, organizing one of the most interesting and variegated polyphonies. The curator, Alessandro Melis, convinced that the convergence of multiple subjects, apparently not directly connected to architecture, constitutes the essential, vital breeding ground of innovative and concrete ideas, in the role of hospitable '*convener*' (Sarkis, 2021), has extended with great openness his table. Referring to mechanisms of resilience observed in biology and neuroscience, such as '*adaptation*' and '*associative thinking*' (Medas et al., 2021), mechanisms that, going beyond deterministic evolutionary logics, offer the ability to adapt and survive in extremely difficult and unexpected situations, thinking about architecture more as ecology than as artifice and rejecting its isolated autonomy, he proposed the trans-laterality of a wide range of connections, to reach flexible variability of suggestions suitable to the most disparate and unpredictable circumstances.

Creativity and diversity of non-systemic thinking have inspired a path that, challenging deterministic aesthetic and cultural homogeneity, declined into 14 thematic sections, related to the crisis of natural and built environment, health, social pressure, creativity and ecology, aimed through cross-pollination and synergy of contributions, as diversified as possible, to lead toward a resilience, which could also be simply called '*self-preservation*'. The journey, privileging experience and immersion, has favored expressive forms linked to creative and industrial arts, as graphic novels and cyberpunk culture, new media that, even if often underestimated, have repeatedly proved to be able to predict contradictions and paradoxes of our contemporary unsustainable cities and environments. The curator, according to the visionary anticipations of the '60s movements of radical groups about the necessity to re-write the elitist academic language, expanding its means more in keeping with a changing and dynamic context, convinced that architecture should not be based on the easy consensus of inertial parts of society, has resorted to the more accessible contents and enormous communicative potential of the latest digital media, in order to involve and interact with a broader audience, especially the younger generation.

Art can often be an essential tool for bonding together various aspects of meditation and debate. In this regard, an artist who has always dedicated his research to the contemplation of our architectural and urban legacy was invited. The foresight of Giacomo Costa's photographic language completes this hybrid ecosystem of microcosms, responding to the 'Resilient Community' (Medas et al., 2021) with four monumental works, from the 'Elements' series, displayed along a 20 m wall. Devastating visions of futuristic megalopolises, "*anthill cities...grown...like cells...replicated out of control*" (Beatrice & Costa, 2020, p. 242), on the verge of extreme, apocalyptic disasters, caused by the four natural elements, disquiet us. The man appears completely absent but his violent footprint dramatically screams in the chilling and crumbling scenarios of these crowded agglomerates, deprived of any liaison with greenery and nature.

Costa's research stems from the urgency to reflect on the meaning of the interaction between man and the environment. His powerful and provocative works focus their narrative on the urban context, as primary subject, like an occasion of considerations about man's conducts and attitudes that have affected a balanced relation with nature and also personal relationships inside human society itself. Dark, impersonal modular elements obsessively repeated, beautiful and aberrant towers, soaring toward the conquest of the sky, emphasize an urban anti-ethics, triumphalistic behavioral patterns that do not admit the humility of recognizing limits and failures.

From an initial traditional photography, the artist moved to sophisticated digital techniques that, manipulated by his great ability, allowed him, fond of science fiction, to become the demiurge of his futuristic perspectives, crafted with hyper-realistic three-dimensional force. The strong graphic impact of his over-sized creations emotionally involves, captures and seduces, pushing us to reconsider a progress that has perhaps altered in a too disharmonious way the vital relationship that should exist between man and his natural context. The complex issues analyzed along the Pavilion's path, magnified and transmitted with the provocative immediacy of the visual artistic, creative voice of metaphorical representations, raise awareness and doubts.

The looming, intimidating threat of these environmental collapses, as in other cases gigantic dams in the act to explode, or fearful, irrepressible rebellions and revenge from the indomitable natural counterpart, should convince to abandon anthropocentric conducts, arrogant, unjustified dominant processes of colonization, accepting to be part of a multi-species coexistence, based on respect, equity and inclusiveness. Costa's "powerful visions"—as Norman Foster, enthusiastic collector of his creations, wrote about him—"*remind us above all of the fragility of our built world and the civic premises that have so far underpinned it.*" (Beatrice et al., 2008, Preface).

Venice Architecture Biennale 2023

The following 18th International Architecture Exhibition reiterates and drastically amplifies the emphasis on openness and diversity. The curator, Lesley Lokko, champion of equity and inclusion, the first black architect to lead the event, addresses the urgency to change conversation globally, and democratize architecture, re-shaping the narrative and pushing the boundaries of the discipline. Her continent of origin, Africa, at the heart of the festival for the first time, offers the chance to re-consider and understand mistakes, opening to a multiplicity of voices that have been silenced for long time. The prefix '*re-*', placed upstream of the program, promotes recurrently the power to re-forge a new vision of the world, through a humble approach animated by willingness to objectivity and dialectic complementarity.

"*New audiences are also emerging, hungry for different narratives, different tools, and different languages of space, form, and place*" (Lokko, 2021) and this edition, the youngest ever, with participants aged around 40, with 50/50 gender balance, provides them the unique opportunity to "*put forward ambitious and creative ideas that help us imagine a more equitable and optimistic future in common*" (Lokko, 2021).

In a '*laboratory*', conceived not as a place of conventional scientific experimentation, but as a "*messy, passionate, argumentative space of deeply human enquiry*" (Lokko, 2023a, 2023b), the '*architects/practitioners*', invested with new, expanded roles of the practice, invited to examine issues that will dominate the next century, focused on sustainability, such as decarbonization and decolonization, ways of building less exploitative of people and nature, share the stage, listening and speaking with a rich field of creative disciplines, promoting attunement and igniting dialogue with the public.

The participants, sole or individual practices of five people or fewer, re-visit the past through a decolonialist lens to trace a possible more equitable, sustainable future, presenting collective research works that reveal different versions from the official ones, completing the process with events, lectures, panel discussions, videos, and performances.

Lokko challenges to re-think scarcity, not as a lack of something but as an abundance of imagination. By connecting memories and experiences, utilizing the potent tool of imagination, places become more resonant and inclusive, and the hope is to provoke "*the audience to think differently and more empathetically*" (Lokko, 2023a, 2023b).

Navigating through the disconcerting panorama of our Anthropocene Age, the water crisis captures attention as one of the most recurrent topics, analyzed in its multifaceted complexity and extreme events, trying to reverse a violent trend, finding nature-based solutions of adaptation from ruinous human coercions.

The Georgian Pavilion, presented by the Tbilisi Architecture Biennial group, tells the story of an artificially altered settlement, a precise case study that from a distant past intends to dwell on the true cost of the '*development*'. It expands the reflection and induces a debate, according to a future, more global perspective, on the exploitation of hydro resources. With the dark mud collected from the banks of the Zhinvali hydroelectric basin, the first power station launched in Georgia, built in 1985 in the Dusheti region for the production of electricity for the capital, bricks were crafted for the Pavilion's installation, 'january february march' (Gurgenidze et al., 2023). The installation reconstructs an absence, the empty mass of one of the artifacts discovered during the research, the vaulted volume of the 12th-century Church of Jvaripatiosani which, submerged by the hydroelectric basin's waters, reappears, with clear allusion of the title, during the months of drought, when Tbilisi consumes the most water.

The plant, 11.5 km^2, still substantial in generating electricity and water supply, has however caused the complete flood of the Zhinvali village, forcing the entire local population to migrate. Eneolithic age settlements, significant examples of cultural heritage, have been destroyed, while the Aragvi River's ecosystem no longer exists.

Metaphoric photographs accompany the performance held in scene, echoing the impact of human interventions on the environment through the juxtaposition of dead and living nature representations, as the desperate gesture of a desiccated tree which emerges skeletal from the idyllic vastness of the water mantle filling the entire reservoir. A publication compiled with the contribution of experts in various fields, video art and documentary materials, giving voices to the protests from the Georgian population against the possible implementation of another hydroelectric plant, displays a plural vision along four rooms. The narrative of the events, entrusted to the visual and sensorial stimulation of the artistic forms, activates meaningful spatial experiential involvement, establishing emotional links with time, and memory of the area, seeking consciousness and urging an exchange of views about this dialectical relationship between man-made and natural forces.

Georgia once had the enviable primacy of energy autonomy from renewable sources, and its rich, diverse water patrimony, approached with great humbleness and sacred respect, has always been a vital part of its ecosystem, economy, agriculture, and culture. Water was a common good, and people used it, in harmonious coexistence, with mills, with irrigation, without compromising its natural course. Colonialism and the exploitation of natural resources, as this transformation of the environment into '*economic value*' are at the heart of the project. Many questions arise, considering the radical, irreversible modifications introduced, both in Soviet times and today, by these large infrastructures that, with the privatization of state assets, have transformed a common wealth into a source of income for a privileged few. Questions about who benefits from transforming nature into a tradable commodity and who bears the costs when such projects lead to negative consequences inflate this inquiry and trigger exchange between the laboratory of practitioners and the audience.

References

Beatrice, L., & Costa, G. (2020). Conversation between Giacomo Costa and Luca Beatrice. In *Giacomo Costa, a helpful guide to nowhere* (p. 242). Damiani Limited.

Beatrice, L., Costa, G., & Foster, N. (2008). In *Giacomo Costa: The chronicles of time* (Preface). Damiani Limited.

Gurgenidze, T., Nemsadze, O., Pasuri, E., Shukakidze, G., & Tornieri, S. (2023). *January February March*. Tbilisi Architecture Biennial.

Lokko, L. (2021, December 14). Lesley Lokko appointed Curator of the Biennale Architettura 2023. *La Biennale di Venezia*. https://www.labiennale.org/en/news/lesley-lokko-appointed-curator-biennale-architettura-2023

Lokko, L. (2023a, May 17). Africa a "powerful place from which to examine the issues that will dominate the next century" says Venice curator Lesley Lokko. *Dezeen*. https://www.dezeen.com/2023/05/17/venice-curator-lesley-lokko-interview/

Lokko, L. (2023b, April 20). Lesley Lokko on the 2023 Venice Architecture Biennale: "I hope it provokes the audience to think differently and more empathetically." *ArchDaily*. https://www.archdaily.com/999671/lesley-lokko-on-the-2023-venice-architecture-biennale-i-hope-it-provokes-the-audience-to-think-differently-and-more-empathetically

Medas, B., Melis, A., & Pievani, T. (2021). Architectonical exaptation and creative serendipity. In A. Melis (Ed.), *Architectural exaptation: Catalogo del Padiglione Italia. Comunità Resilienti alla Biennale Architettura 2021* (Vol. 01a). D Editore.

Sarkis, H. (2021, May 6). Biennale Architettura 2021: How will we live together? *La Biennale di Venezia*. https://www.labiennale.org/en/news/biennale-architettura-2021-how-will-we-live-together

Chapter 17
Advocating Ecology Through Musical Experimentation and Performing Arts

Enrico Lain, Saverio Massaro, and Luca Lagash Saporiti

Abstract This chapter investigates the artistic process UUUUDUM—*Unorthodox Uncontaminated Universe: Use, Develop and Unlock Music*, hereafter referred to as [U], based on musical and theatrical performance, and mixed media, proposed by the OP Collective, founded by MOG Morgana Orsetta Ghini, Luca Lagash Saporiti Thomas Böhm and Paolo Grigolli. The present investigation is motivated by two primary factors. Firstly, it stems from the collaboration between the OP Collective and The Non-Humans' Embassy in Favara (AG) for the development of [U]. Secondly, from a methodological point of view, it delves into insights on agency derived from the study of performance within the musical sphere. The work done by [U] proposes to amplify the effects of natural heritage agency, so as to conceptually transform it from an *object of protection* to a *cultural agent*. Through sculptural and musical artistic performances, strategies for the enhancement of large portions of the landscape are proposed. The enhancement of their agency potential, primarily through musical practices (exploring sense via listening to and recording sound objects), proposes an alternative ecological perspective characterized by a project-oriented approach. [U]'s work is made explicit in relation to the conceptual framework already explored with the Non-Humans' Embassy, with further insights into the performing arts elaborated through the contribution of experimental music. This extended conceptual framework will be based on three cores. The first concerns musical experimentation, which will be presented as an interactive method for exploring the real. For it, some parallels between concepts related to the *event* and those derived from meaning will be exposed. The second core concerns *listening*, which will be described as a moment of composing the possible. The third core will offer some reflections on the concept of *instrument* and its ability to deepen the concept of agency. Finally, the synergy

E. Lain (✉)
Selvazzano Dentro, PD, Italy
e-mail: enrico.lain@blu143.it

S. Massaro
Altamura, BA, Italy
e-mail: saveriomassaro86@gmail.com

L. L. Saporiti
Milan, MI, Italy
e-mail: lucalagash@googlemail.com

A. Melis and M. Pelle (eds.), *Design for Performative Arts Spaces*,
https://doi.org/10.1007/978-3-031-98215-6_17

between [U] and NHE is framed in order to promote challenges and opportunities for Italian and European cultural centers.

Keywords UUUUDUM · Unorthodox Uncontaminated Universe · Use, develop and unlock music · OP collective · Artistic process · Musical performance · Theatrical performance · Mixed media · Collaboration · Non-Humans' Embassy · Agency · Natural heritage · Cultural agent · Sculptural performances · Landscape enhancement · Musical practices · Ecological perspective · Project-oriented approach · Conceptual framework · Musical experimentation · Interactive method · Listening · Composing the possible · Concept of instrument · Synergy · Cultural centers · Challenges and opportunities

Methods of Inquiry and Exploration of the Real in the Case of UUUUDUM

This study examines the relationship between the Non-Humans' Embassy, hereafter referred to as NHE, installed within Palazzo Miccichè in Favara (Agrigento, IT), and the artistic process *UUUUDUM—Unorthodox Uncontaminated Universe: Use, Develop and Unlock Music*, hereafter referred to as [U], conceived by Collettivo OP (founded by artists MOG-Morgana Orsetta Ghini, Alessandro Cremonesi, and Luca Saporiti Lagash) and based on musical and theatrical performances and mixed media.

This interaction, promoted by the independent cultural production center FARM Cultural Park in Favara, has facilitated the establishment of a partnership aimed at further developing NHE[1] as a cultural outpost focused on enhancing the agency of non-humans. This text aims to describe some of the operational insights that emerged from the encounter with [U], thanks to its musical approach to reality.

[U] is an artistic initiative born from collaboration with the HTW Berlin technology hub, spanning from October 2023 to February 2024. It builds on the groundwork laid by "Uno Di Un Milione" started in 2019 with Accademia Teatro alla Scala di Milano, Val Di Sole Tourism Board, and UN Mountain. This project encompasses the Presena glacier, the summit of Pejo 3000, and the springs of Stelvio National Park.

[U] employs musical experimentation as its conceptual framework, developing in three stages. The first one involves sculptor MOG creating ten symbolic seating components representing environmental issues.[2] Stage two features ten thematic events broadcast online. Technical partners craft seating using reclaimed materials

[1] See more: https://www.farmculturalpark.com/ambasciata-dei-non-umani.

[2] The issues are: soil consumption, global average temperature, migrations, sea levels, water scarcity, airborne particulate concentration, atmospheric CO_2 concentration, species extinction, habitat loss, and green jobs.

and advanced technologies.[3] Users interacting online through sound avatars generate material for sonorizations, which will accompany the third phase: the installation of the complete work, comprising all ten seating elements. This will constitute a seminal and ritual outpost, where guests can focus on the social dimension of sustainability (fundamental rights, gender equality, and diversity), sharing innovative ideas and works promoting recycling and renewable energies.

The sound avatars for social media interactions have a specific origin: each is a symphonic-derived sound object, stemming from the decomposition of the piece recorded in 2021 by the Accademia Teatro alla Scala di Milano and the sound production developed during research workshops conducted with the technological hub of HTW Berlin University between October 2023 and February 2024. This recording took place at altitude, atop the Presena glacier in Trento, from which the symphony gathered sonic material to compose into musical material. The soundtrack—orchestrated by Lagash himself—provides an extra-local and extra-material field of interaction in relation to the sculptural component. With the technological support of PPK Innovation, the activation of musical fragments is underway, composed of thousands of sound avatars assigned to users upon registration on the web platform[4] of [U].

Conceptually, the [U] process can be seen as a strategy to enhance and amplify the agency of natural heritage. The enhancement of this capacity to act, primarily exercised through musical experimentation,[5] suggests a design thinking approach where the form of the work emerges as the outcome, rather than the premise, of field research.

From its inception in the Trentino Dolomites, [U] manifests as a poetic experience grounded in rigorous empirical methods. The focus of its research is not an abstractly *realistic* representation, but rather a *concrete* reality. In the context of our predominantly visual culture, this distinction can be conceptualized as a shift from *looking* to *listening*. From a philosophical standpoint, this approach aligns with the conceptual framework of *actor-network* theory, which posits that reality is not an inherent, static entity, but rather emerges concretely through interactions (Law & Mol, 1995, p. 277). Given [U]'s focus on redistributing agency, it can be conceptualized as an actor-network itself. As this study will demonstrate, [U] displays a distinct sensitivity to heterogeneity and possesses the capacity for ongoing self-redefinition (Callon, 1987, p. 93).

Methodologically, [U]'s empirical approach is deeply rooted in musical experimentation. This innovative method provides access to concrete reality by promoting a renewed engagement with objects, which are conceptualized as contested assemblages of multiple, often conflicting, agencies (Latour & Yaneva, 2017, p. 104).

This object-oriented approach is not confined to individual elements but primarily concerns the potential *compositions of elements* for which human interaction is

[3] Through partnerships with a number of companies, such as Nugae, Ricehouse, and Arche 3D.

[4] http://uuuudum.collettivoop.com.

[5] The periphrasis "musical experimentation" is used in the text with a direct reference to Scheaffer's "musical," which he defines as "a frontier land where the Arts, like the Sciences, have to be involved" (Schaeffer, 1966, p. 12).

both foundational and limiting. This is exemplified by the concept of landscape. As Jakob argues, landscape design and the tradition of landscape painting have prioritized subjective interpretation over objective reality (Jakob, 2009, pp. 83–84). The challenge, therefore, is to find a new balance—both cognitively and compositionally—between subjective and objective dimensions.

From an epistemological perspective, the distinctive feature of [U] lies in its capacity to bring the conflictual nature of objects and landscape to the forefront of auditory attention, while attempting to propose novel compositions capable of interweaving their visible and sonic dimensions. For [U], the sonic dimension represents an extremely concrete stratum of reality. This perspective, which might appear counterintuitive, aligns with the theoretical framework proposed by Schaeffer. In his 'Traité des objets musicaux' Schaeffer considers musical experimentation as the set of processes that enable a comparison between the physical component and the experiential, subjective component of the sonic *event* (Schaeffer, 1966, p. 119).

The term "event" is closely related to empiricism: events are ephemeral components that are difficult to define and challenge the logical structures of disciplines (Žižek, 2014, p. 20). Musical experimentation does not intend to understand the event, but, according to Schaeffer, aims to explore it through the study of the relationship between it and its listener (Schaeffer, 1966, p. 115).

This dual relationship enables a mediated understanding between the listener and the sound objects, specifically the recordings of events explored by [U]. These recordings serve as the material to be assembled into soundscapes that Lagash will later compose for MOG's ten sculptural sittings. These, through soundscapes, will express the traces of the original events that produced the sounds, encapsulated in their sound spectrum. These traces represent the essence of the sound objects and offer the listener a type of knowledge that hasn't yet been reduced to simple meaning. Indeed, the event itself is the very sense[6] (Deleuze, 1969, p. 27).

The empirical setting of musical experimentation allows the event to be explored as a mediator (Latour, 2005, p. 72) in the ongoing flow between sense and signification. The collection of musical events, situated between the world and the listener, creates a fragile *surface of sense*. This surface metaphorically connects the series of signifiers (the musical languages) with the series of meanings (the world around us).

Schaeffer defines this surface as a gap between acoustic components and music, proposing a hybrid discipline that combines the scientific description of sounds with the art of listening (Schaeffer, 1966, p. 12). However, caution is necessary because, like the surface of a lake, sense is fragile: it can provide resistance yet also allow things to sink into deeper depths (Deleuze, 1969, p. 115).

In [U], one can envision the surface of sense akin to the surface of water, blending philosophical inquiry and musical experimentation into a unified symbolic synthesis. The flow of water, originating from the landscape of the Presena glacier and the springs of the Stelvio National Park, moves through vast landscapes and territories,

[6] Here the words "sense" and "meaning" are used with a subtle distinction, as Deleuze did in his *The Logic of Sense* (see References).

acting as a mediator between the visible and the invisible. This condition is eloquently captured in a text by Alessandro Cremonesi[7]:

> If you love poverty and hiddenness then you are like me. If you prefer humility you are still like me who is always sliding downward, lower than everyone, and that is the only way I know how to make things bear fruit.

The surface of water, much like sense, remains neutral as it flows through human-made spaces, touching plants and animals (Deleuze, 1969, p. 100). Sense, like water, is essential. It enriches culture by existing before the formulation of meaning and propositions (ibid., p. 115), enabling questions to be posed freely.

Similarly, in [U], the exploration of the work's sense becomes the central and connecting element. The ultimate meaning is continuously deferred, awaiting the conclusion of the process. This occurs when all agents, having transformed into sound simulacra, can converge onto a cohesive surface of sense, spread across the territories of the intricate natural heritage awaiting interpretation.

The Role of Listening and Musical Instruments

The event is elusive because it defies tracing its causes or being fully present; it evades the present by transcending the constraints of any fixed state (Deleuze, 1969, p. 135). Always awaiting representation by the subject, the event is inherently disruptive. This disruptive nature has been explored in subjective syntheses, like Schechner's performance theory, which links the event to a subjective, memorial-like ritual, with the aim to discover and restore what has already been stored in social memory (Schechner, 1984, p. 19). Therefore performance is often constrained to recall existing knowledge (Stiegler, 2015, p. 42). This limitation was evident in the operational aspects of NHE performances at FARM (Favara, August 2023)[8] and Teatro de LiNUTILE (Padua, September 2023). In those performances language still played an excessive role in relation to bodies and their choreographic expression. In addition, monologues encapsulated identities that were far too delineated, overly significant, imposing themselves as ballasts over the surface of sense. To truly capture the event, where past and future converge (Deleuze, 1969, p. 76), performance must transcend mere ritual, intertwining the memorial with the potential, the experimental with the ideal (Deleuze, 1969, p. 26). The resulting openness is undifferentiated, without distinctions between the specific and the general (ibid., p. 53). However, in this neutral flow, it remains essential to be able to determine states of suspension.

Within musical experimentation, *listening* is defined as this fragile suspension of the agential flow that precedes the formation of meaning. In listening, memorial time, intentional orientation, nostalgia and desire coexist as conditions of sense (ibid.,

---

[7] Cremonesi is also the author of the Poetic Manifesto of [U], which can be read at the link: https://uuuudum.collettivoop.com/manifesto-poetico/.

[8] Watch the video: https://youtu.be/Oy7AewJE9X0.

p. 72). Listening, close to both hearing and understanding, operates in an indistinct, inorganic, nonhierarchical way: no ears are needed to listen, it is the *whole body* that listens to other bodies.

In listening, there is no action but only a one-time orientation (Hersh, 1990, pp. 70 and 71). Thanks to it, in musical experimentation, the problematic, the possible and the *error* can coexist (Deleuze, p. 104), without yet being subject to verification. For these reasons, in electroacoustic music, even error and failure could become new sound objects for new compositions (Cascone, 2000, p. 13). This is an extension of the sound spectrum, which can also include transit components (Schaeffer, 1966, p. 151), dissonances and noise, with the aim of implementing the very conditions of music making.

These components coexist with pure sound, and the sensitivity of musical performance allows them to be detected, showing a much more articulated reality than is visible. Deleuze defines the set of these components as a *problematic field* (Deleuze, 1969, p. 56), to be understood not as an imperfection in the process of knowledge, but as that which allows the very genesis of solutions to be organized (ibid., p. 55).

The poetic and productive potentialities of [U], in its operation in this problematic field, can be described through some reflections on the *instrumental.* The instrumental is different from the mechanical because of the former's peculiar disposition toward the unfinished, especially in the creative act. It in fact imagines the product while simultaneously finding the tools for its realization (Brock, 1975, p. 79). While the machine is a trap set for the natural and generates a fragmented temporality (Lyotard, 1975, p. 105), the tool, on the other hand, has a mediating role along the surface of sense. In the *musical* instrument, this mediation is perhaps more extensive, as it continually acts between opposites: between permanence and variation (Schaeffer, 1966, p. 31), between the physical world and the aesthetic imagination (ibid., p. 35). These mediating characteristics become more evident in experimental music, in which composition explores musical structures through the study of sound objects, searching for elements with modes of existence that are prone to coexist in composition due to their own imperfection. Imperfection is shown as a premise of composition. In this, sound objects and technical objects (Simondon, 1958, p. 50) present similar ontologies. For both types of objects, materializing into assemblages is a process that, while involving the subject, is also based on their imperfection (ibid., p 71) and their mutual compatibility. Musical experimentation, thanks to the practice of listening and the mediation of the instrument, makes it possible to operate within this reciprocal and imperfect openness, at that stage of time orientation in which agency is still potential and has not yet produced detectable differences.

The Generative Process of UUUUDUM

The exploratory and compositional process of [U] operates as a generative—rather than productive—system capable of highlighting relationships of interdependence between objects and fostering a redistribution of agency between them and humans (Fig. 17.1).

Thanks to the musical experimentation within the process of [U], the description of the extinction of the Presena glacier takes place through the recording of its sound events, which trace the flow of water. The symphonic piece that recomposes the sound objects represents an initial determination of sense, thus offering a possible account of the glacier's past and future. Subsequently, through the deconstruction of the architecture of the symphonic piece, each resulting sound object, once split, transmitted and shared as a sound avatar, is able to disseminate its own components, after making its origin and sound spectrum evident, so as to simplify its understanding.

The orchestral performance was recorded with the participation of the symphony orchestra of the Accademia Teatro alla Scala in Milan in October 2021. To obviate the restrictions for the pandemic, new positions among the musicians and unprecedented microphone shooting modalities were experimented with. Technically, the first phase of composition and orchestration was carried out inside the Presena glacier, using instruments built from ice inside the Ice Dome Theatre—an igloo theater made of ice.

The completed symphonic work was then deconstructed and aggregated to the production of electronic and natural sounds. The process took place within the HTW technology hub in Berlin, thanks to the use of softwares like Pro Tools and Live Ableton and to the skills of digital developer engineers; the latter have developed the necessary codes to ensure, in the production of the avatars, the assignments between

Fig. 17.1 The summit of Pejo 3000, in the heart of the Ortles-Cevedale mountain group, in Val di Sole, Trentino

people and sound objects. In this way, each sound avatar itself becomes part of the soundscape works to be installed in the ten events distributed along the Italian territory.

Also within HTW Berlin, the sculptural pieces were reworked into as many digital devices for AR and VR environments, with which to interpret climate themes within immersive 360FullDome applications.

Integrating Environmental Stewardship in Cultural Initiatives

Events are essential not only for empirical knowledge but also for grasping the distribution of agency. Following Latour's irreductionist approach, the world isn't structured by substances but rather forms a network of connections that converge into nodes. These nodes represent unique and irreducible events where the forces of actors come together and condense.

This is crucial to reveal the interpretative limitations of the world that modern culture has attempted to hide or deny, and to acknowledge the latent agencies in artifacts and technical objects with programmed actions capable of interacting with their environment, which are increasingly present in our surroundings (Rammert, 2008, p. 74).

In the case of [U], creating sound avatars transforms recording into meaningful action, as these artifacts actively seek interactions that generate critical differences and constitute significant meanings beyond the initial act. The avatars multiply and redistribute the agency of the Presena glacier.

Artifacts and interactions animate [U]'s system and its exchange flows with the NHE to build a network of mutual influence that could eventually function as a cohesive whole.

In the current historical phase dominated by graphic-visual culture and AI linguistic models, the practices of listening and musical experimentation emerge as critical methodologies for re-evaluating existing design paradigms.

The characteristics embodied by [U] and its performative modes render it a cohesive and complementary process, whose outcomes can form a shared legacy to be enhanced through the NHE. The interaction between NHE and [U] is framed in actancial terms, exploring which elements contribute to the realization of a particular event, the conflicts they may generate, the relationships they establish, how they associate in various forms, and the assemblages they produce.

The research and experimentation carried out thus far are envisioned to become a common asset among the transformative initiatives across new cultural centers in Italy and Europe. These centers are multifunctional spaces for cultural production or consumption (Niessen, 2019), now including makerspaces, experimental libraries, regenerated spaces, community hubs, and artist residencies (ibidem).

To advance a profoundly ecological agenda, cultural centers could pursue two strategic directions: firstly, recognizing natural landmarks such as the Presena glacier and the Pejo 3000 peak as cultural centers in their own right, empowering them to champion environmental stewardship. Secondly, adopting innovative models such as becoming "Embassies," akin to the transformative example set by FARM in Favara, to redefine their roles through expanded programs and engagements.

In conclusion, by embracing such approaches, cultural centers can play a pivotal role in advocating for and implementing environmental agendas, thereby contributing to highlight alternative modes of coexistence and cohabitation.

References

Brock, B. (1975). Immacolata concezione e macchina celibe. In H. Szeemann (Ed.), A. Bacigalupo, & A. Fiume, (Trad.), *Le macchine celibi* (1989). Electa.

Callon, M. (1987). Society in the making: The study of technology as a tool for sociological analysis. In W. E. Bijker, T. P. Hughes, & T. F. Pinch (Eds.), *The social construction of technological systems—New directions in the sociology and history of technology* (1997 ed.). The MIT Press.

Cascone, K. (2000). The aesthetics of failure: Post digital tendencies in contemporary computer music. *Computer Music Journal, 24*(4), 12–18. https://doi.org/10.1162/014892600559489

Deleuze, G. (1969). *Logica del senso* (2022 ed.) (M. De Stefanis, Trad.). Feltrinelli.

Hersh, J. (1990). *Tempo e Musica* (2009 ed.) (R. Guccinelli, Trad.). Baldini Castoldi Dalai editore.

Jackob, M. (2009). *Il paesaggio* (2020th ed.). Il Mulino.

Latour, B. (2005). *Riassemblare il sociale* (2022 ed.) (C. Gibella, F. Cogliati, L. Novelli, M. Morlacchi, M. Alduino, Eds.; D. Caristina, Trad.) Milano: Meltemi Press Srl.

Latour, B., & Yaneva, A. (2017, ottobre 01). «Give me a gun and I will make all buildings move»: An ANT's view of architecture. *Ardeth*, 103–111. Retrieved June 25, 2024, from http://journals.openedition.org/ardeth/991

Law, J., & Mol, A. (1995). Notes on materiality and sociality. *The Sociological Review, 43*(2), 274–294.

Lyotard, J. F. (1975). In cui si considerano certe pareti come elementi potenzialmente celibi di alcune macchine semplici. In Szeemann, H. (Ed.), A. Bacigalupo, & A. Fiume (Trad.), *Le macchine celibi.* (1989). Electa.

Niessen, B. (2019, October 17). Cosa sono i nuovi centri culturali, l'avanguardia della trasformazione culturale. *CheFare.* https://che-fare.com/almanacco/politiche/comunita/cosa-sono-nuovi-centri-culturali/

Rammert, W. (2008). Where the action is: Distributed agency between humans, machines, and programs. In U. Seifert, J. H. Kim, & A. Moore (Eds.), *Paradoxes of interactivity: Perspectives for media theory, human-computer interaction, and artistic investigations* (pp. 63–91).

Schaeffer, P. (1966). Treatise on musical objects—An essay across disciplines (2017 ed.) (C. North, & J. Dack, Trad.). University of California Press.

Schechner, R. (1984). *La teoria della performance (1970–1983)* (V. Valentini, Ed.). Bulzoni Editore Srl.

Simondon, G. (1958–2012). *On the mode of existence of technical objects* (2017 ed.) (C. Malaspina, & J. Rogove, Trad.). Univocal Publishing.

Stiegler, B. (2015). *Platone digitale - per una filosofia della rete* (P. Vignola, F. Vitale, A cura di, & P. Vignola, Trad.). Mimesis.

Žižek, S. (2014). *Evento* (2015 ed.) (E. Acotto, Trans.). Milano: UTET.

Chapter 18
The City as an Event. The Performative Architecture of Seoul's Commercial Spaces

Dario Pedrabissi

Abstract This chapter investigates the commercial districts of Seoul, where dense clusters of neon-lit signage transform urban space into a spontaneous, performative environment. These areas, dominated by vibrant and chaotic visual landscapes, challenge traditional notions of architecture and public space by presenting signage as a form of urban scenography. The research is timely and relevant as digital technology continues to undermine physical public interaction, rendering many global commercial zones increasingly obsolete. In contrast, Seoul's districts remain dynamic hubs of activity, not despite their visual noise, but arguably because of it. The city's signage-saturated façades offer a contemporary model for urban vitality and social engagement, where consumerism and spectacle intersect. Employing a methodology of participant observation, the study draws on walk-throughs, video recordings, photographs, sketches, and field notes to capture the lived experience of these spaces. This ethnographic approach foregrounds the sensory and spatial impacts of signage in specific Seoul neighbourhoods, highlighting their spontaneous and uncurated character. Findings reveal that commercial signage in Seoul transcends mere advertisement, functioning as a performative layer over standardised architecture. These signs—ranging from oversized octopus sculptures to immersive neon displays—generate a participatory and theatrical urban experience. Far from curated or architect-designed, these environments are shaped by competitive individual expression and rapid turnover, resulting in a dynamic visual culture akin to a collective artwork. The study argues that these commercial spaces exemplify a form of urban vitality rooted in spontaneity, consumer spectacle, and symbolic overload, offering new perspectives on the role of public space in the digital age.

Keywords Commercial signage · Performative urbanism · Informality · Public space · Consumer spectacle · Seoul cityscape · Visual culture · Ethnographic methodology · Digital urbanism

D. Pedrabissi (✉)
School of Architecture, Art and Design (AAD), University of Portsmouth, Portsmouth, UK
e-mail: dario.pedrabissi@port.ac.uk

A. Melis and M. Pelle (eds.), *Design for Performative Arts Spaces*,
https://doi.org/10.1007/978-3-031-98215-6_18

Introduction

Commercial signage in East Asia become a landmark of the hectic public life in the vibrant oriental megacities. In South Korea, commercial areas with signage packed in a medley of letters and figures enhanced by coloured neon are omnipresent. This exuberant and creative visual representation of interior function simultaneously generates chaos and a dynamic public space. This chapter focuses on Seoul and its commercial areas, seen as performative spaces generated by the chaotic juxtaposition of commercial signs and their impact on the public space and its use.

Public space plays a crucial role in the modern city, mainly because the digital world dominates social life and deteriorates the role of physical space. The commercial districts have also suffered from this process, and many shops closed. The result is that people's interaction is reduced, and the public space gradually loses its innate quality to attract diverse groups of individuals. However, the Korean commercial districts are a positive example of public space, although very chaotic and hectic. Also, they are intensely focused on consumption and less on a free interpretation of space as in traditional public spaces like the *piazza*. However, can the architecture shaped by exuberant commercial billboards be considered a spontaneous performative event at the scale of the city? Moreover, what is their impact on the public space, and how does it affect the users?

The result reported in this short chapter regards the comprehensive understanding of the commercial areas in Seoul, as the accidental signage exhibition is composed of different elements. The public space is the actual performative space, where people experience this overwhelming impact of light and sound. The architecture behind the signage is a standard multistorey building supporting the advertisement. The commercial signage is the outer layer of the building and gives the business owner individual expression. The architect has no role in the appearance of the building. Thus, the individuals orchestrate the whole, as no curator is behind this process. A spontaneous scenography is exhibited for the consumer society.

Signage as Symbols in Architecture

Symbols in architecture have been used for millennia. The need to communicate through architectural elements is full of examples, such as the hieroglyph on obelisks in ancient Egypt, the more abstract symbology of garlands and eggs on the trabeation of ancient Greece, the medieval frescoes, or, again, the widespread adoption of sculptural works in the Mannerist facades. Modern architecture then banned any decoration on the buildings. Still, the human eye is eager to contemplate details and curious to look at symbols to establish a formal interaction with the surrounding environment. Since the 1990s, contemporary museums have played—once more—a crucial role in the city, becoming themselves the symbol of culture and a hedonistic city attraction.

They continue their traditional mission of preserving and showing objects. Simultaneously, they are a sculptural manifestation to be observed and experienced, glorifying their architectural appearance and rediscovering their communicative function (Celant, 2006).

Communicating is the main mission of an exhibition. Therefore, the museum can tell its story through both architectural form and exhibition design. For the latter, the critical aspect of a good exhibition design is to create good storytelling to navigate the different spaces through groups of artefacts. Furthermore, the recent trend of 'experience design' in exhibitions that emerged from the commercial sector is directed to engage the consumer with various technologies to immerse the senses into the exhibition (Dernie, 2006, p. 23). A similar immersive experience is also performed in commercial districts, where billboards, neon lights and new technologies attract the consumer society and transform their relationship. This is a phenomenon anticipated by Debord (2014) in The Society of the Spectacle, paragraph 4, where he states, "The spectacle is not a collection of images; it is a social relation between people that is mediated by images".

The symbolic value of architecture is analysed through the groundbreaking books of Robert Venturi, Denise Scott Brown, and Steven Izenour. With this publication, the authors have introduced new methods to study a modern city that cannot be referenced to any traditional, well-established environment, such as the Italian *piazza,* claiming that architects are ill-equipped to analyse a modern urban condition characterised by an apparent disorder. In their study on analysing the massive billboards that dominate the Las Vegas Strip, they describe the commercial advertisements: "Signs in Las Vegas use mixed media – words, pictures and sculpture – to persuade and inform. A sign is, contradictorily, for day and night. The same sign works as a polychrome sculpture in the sun and as a black silhouette against the sun; at night it is a source of light. It revolves by day and becomes a play of lights at night". (Venturi et al., 1977, p. 52)

Although the vast space between buildings in Las Vegas is the opposite of the Korean environment, characterised by compactness and high-density districts, the analysis of billboards developed by Venturi, Scott Brown and Izenour reveals a new way of interpreting the cityscape. Another pivotal text on American culture of the 1970s is Los Angeles: The Architecture of Four Ecologies by Reyner Banham. For him, commercial symbols are an essential part of the culture of Los Angeles. For a commercial operator, the best option would be to acquire a standard building, or even a disused one, and "impose your commercial personality on it with symbolic garnishes" (Banham, 2009, p. 101).

In contemporary large cities and key commercial areas, we are used to being overwhelmed by massive digital billboards projecting captivating images and brand logos. Famous locations for such an experience are London Piccadilly Circus and New York Times Square. In Korea, the Fourth Screen is a new three-dimensional experience at the SMTown Media Facade at the COEX centre in Gangnam, Seoul. In some cases, media have been integrated into the architecture, not to enhance the building's quality but to be a principal architectural element, driving the whole project. The Tower of Winds by Toyo Ito & Associates is the most famous example, built in Tokyo in 1986. The project was composed of several lights and neon. During the day, the tower had a humble appearance, while during the night, the lights were activated by the surrounding sounds, and the speed of the wind activated the ring neons, resulting in a choreography of lights that was always different.

A much more recent project that uses light for its facade is the Galleria shopping mall in Cheonan, South Korea, designed by UN Studio in 2010. The facade is more dynamic due to the architectural mullions that create visible waves day and night. However, at night, the building reveals all its potential, with the great variety of patterns projected from the double layer of the facade. Still, the Toyo Ito project had an extraordinary effect: creating a direct relationship with the surrounding environment and responding to it with varied lighting conditions.

These cases are contemporary interpretations of the Asian eagerness of scintillating buildings and sparkling lights to interact with the urban environment. Particularly for the Galleria project, the building became a commercial sign, as the facade acts as a giant billboard, not projecting any brand but as the actual symbol of commerce.

Commercial Signage in Korea

Korea did not have a tradition of exuberant signboards. With the advent of foreign influence in the late nineteenth century, large advertisement boards started to dominate the main streets. Furthermore, the Japanese colonisation period (1910–45) influenced Seoul's widespread adoption of commercial signs. In Japan, this culture dates back several centuries. The famous woodblock prints by Katsushika Hokusai, Utagawa Hiroshige, Kitagawa Utamaro, and Keisai Eisen, among others, depicted street scenes between the eighteenth and nineteenth centuries, where signboards, flags, parasols, and other boards on the street were omnipresent in each scene. In Korea, commercial signs were written in Chinese characters to be comprehensible by both the Japanese and the (literate) Koreans. There are images of Seoul in the late 1950s with a mix of signboards written in both Korean and Chinese characters.

Nowadays, the culture of commercial signs is deeply rooted in Korean society. Hence, images of billboards on building facades are intrinsic characteristics of the urban environment. Furthermore, thanks to the ever-growing impact of Korean soft culture abroad, dense and colourful street scenes are part of tourist attractions, such as the Myeongdong neighbourhood. In this regard, the so-called Korean Wave or *Hallyu* is a contemporary cultural phenomenon that started during the 1990s, and that is spread worldwide through the success of K-pop performers such as BTS and Psy's "Gangnam Style", K-dramas, high-quality films that have been awarded in the most prestigious films academies, and fashion designers that are influencing trends also in the western world.

The image of South Korean pop culture is now admired abroad, so articles and documentaries on the Korean Wave are published in different languages. The Seoul Metropolitan Government is boosting tourism, taking advantage of this international popularity. Hence, exhibitions are promoted abroad to strengthen the global recognition of Korea as a cultural hub. In London, a recent exhibition was shown at the Victoria and Albert Museum entitled Hallyu! The Korean Wave, showcasing "…the colourful and dynamic popular culture of South Korea, exploring the makings of the Korean Wave and its global impact on the creative industries of cinema, drama, music, fandom, beauty and fashion." (Victoria & Albert Museum, 2023). At the time of writing, April 2024, an immersive exhibition was shown in London: Delight, Media Art Exhibition, an immersive and interactive installation that conveys "a fresh and creative interpretation of urban city life" (Delight, 2024). The exhibition wants to recreate Seoul's lively urban environment through video projections and interactive technologies. The focus is on the overwhelming representation of commercial signboards, which are the iconic symbols of Korea nowadays. Thus, the signs became icons of Korea and the dynamic urban life of Seoul.

The urban concentration of commercial activities in specific areas was provoked by a condensed accumulation of businesses that competed by creating signboards of excessive dimensions. These districts are less attractive during the day than at night, although many commercial activities are open during the day. However, during the night, the commercial district exalts all its visual features. Furthermore, commercial signs are crucial in the visual features of urban streetscapes (Meikle, 2013, p. 672). Dedicated commercial areas are thus embedded in modern Korean culture. However, this clear subdivision of functions expanded during modernisation, distinguishing functions among residential, working, recreational and circulation sectors. This rational subdivision of functions has flourished in Korea also because of its cultural background. In fact, the Korean rational thinking background comes from the traditional Confucian education and the military regimes that the society experienced under the Japanese occupation and the military regimes from the 1960s to the 80s (Moon, 2005).

The Seoul Metropolitan Government (2008) provided a guidebook to diminish the visual pollution (Blewitt, 2015, p. 181) derived from the large signboards. The commercial areas are a condensed milieu of recreational functions, such as restaurants, clothing and cosmetics shops, bakeries and bars of any kind, where billiards bars and 노래방 (nolaebang or karaoke bar) are among the most popular. Other typical functions are small spas 찜질방 (jjimjilbang), sometimes located underground in a building, Taekwondo martial arts schools, English language schools, and after-school private education institutions called 학원 (hagwon). These buildings also contain more spiritual functions, such as Christian churches and Buddhist temples. Although they are a much smaller version of the buildings, we usually associate them with these functions. Their symbolic representation is made apparent outside with a prefab pinnacle representing a small bell tower for the Christian church and the typical Buddhist icon for the temple. All these activities take place on different floors, adding a vertical circulation departing from the street level, simulating the small alleys departing from the main street of the ancient city.

The Commercial Building as a Scenography

The commercial buildings in these districts are usually between two and five floors and can be used without an elevator, although most have one. The building is a standard design with a side stairwell serving one central space. The facade is unpretentious, as it only needs to support the signboards. Signs often cover the façade comprehensively, so the architectural finishes are optional. The facade must only work as a skin to enclose the interior space. The final aesthetic is defined by the signs that act as the architectural decoration of the commercial building.

Access from the street is usually direct to the stairwell and elevator, without any entry lobby. The signs for each business also continue inside the stairway, with specific graphics related to the company. However, Korean users know where to go even before going to the commercial area, as the restaurants, bars, and other businesses are advertised in the popular blogs in the local Naver internet browser. Thus, social media and blogs recreate online the physical commercial competition visible in the district (Fig. 18.1).

The sign advertising the business of the interior units attracts people from the main route to diverge into a secondary path, which is the vertical circulation provided by the multistorey buildings. Each small commercial building of two to five floors represents small alleys from the main circulation, which is the street level. However, this situation has nothing in common with traditional commercial settings. Only modern technology, which has developed into modern architecture, has allowed this

Fig. 18.1 Commercial district in Seoul. 2022 (Picture from the author)

environment to be created. In such a high-density environment as Seoul, vertical circulation is inevitable.

Thus, if the buildings act as a stage with colourful choreography, the street is where the spectator experiences this spontaneous exhibition of symbols and forms. The visual experience is created by an improvised juxtaposition of different tastes brought together without a single figure curating the overall scene. The public space between buildings is the actual performative route, as the signboards are experienced at the street level. Most districts are mainly pedestrian or allow minimal car circulation, generating a different experience than the Las Vegas example. Thus, a sense of compactness and density is augmented by the tight street size and the height of the buildings, which compress the space even further. The visual experience comes from signs, neons, and three-dimensional figures. The latter represents shapes, such as giant octopuses or tuna, large beer glasses, billiards balls, etc., to communicate and attract customers. Here, the words are superfluous as the three-dimensional symbol intends to dominate the facade of the building. The figure wants to win against the rival with a more obvious declaration.

Thus, the signs work as architectural symbols for our contemporary society, similar to what happened in the ancient buildings, but with consumerism as the primary purpose. In this sense, the public space is contaminated by the surrounding buildings, and the citizens/customers browse around from one sign to the next, looking at the different levels of the buildings and searching for the next ludic experience.

Discussion: The Value of Spontaneous Design

The commercial districts act as a performative space for the consumer culture. They are a pop representation of our hedonistic contemporary lifestyle. When combined with a multitude of words, shapes, and neons, the colourful commercial signs are a visual representation with an accidental artistic value. The final urban scene results from a spontaneous accumulation of objects with a powerful visual effect as the primary goal. However, their overall composition in the single facade of the building and, on a larger scale, in the commercial district is uncontrolled by a single curator as it instead happens in the formal exhibitions in museums and galleries. Competition among different businesses is the only drive for creativity. This informal performance is also temporary as the shops and bars change relatively quickly. New business owners come and decide on the new lettering, colours, sizes, and figures to squeeze their ideas into the already packed building facade.

The main issue is that architecture is denied. The standard building works as a scaffolding for commercial signage, producing visual and sound pollution, which the SMG has tried to control, at least in the newly developed areas. However, this phenomenon produces an environment overcharged with vitality due to its compressed accumulation of functions unfolding at different levels of the buildings. At the street level, the effect of this spontaneous exhibition of signs and symbols on the users—as I experienced through my fieldwork—is exciting, even exhilarating, as there are analogies with amusement parks. The space is soaked with people, sound, lights and movement-generated dynamism. There is a sense of modernity or even a futuristic atmosphere. A 'spectacle' of images with which humans interact.

References

Banham, R. (2009). *Los Angeles: The architecture of four ecologies* (3rd ed.). Univ of California Press.

Blewitt, J. (2015). Visual pollution: Advertising, signage and environmental quality. *Planning Perspectives, 30*(1). https://doi.org/10.1080/02665433.2014.967496

Celant, G. (2006, February). L'immaginario del Museo. *Casabella*, (741), 6–7.

Debord, G. (2014). *Society of the spectacle.* Bread and Circuses.

Delight: Media Art Exhibition (2024, April 6).

Dernie, D. (2006). *Exhibition design.* Laurence.

Meikle, J. L. (2013). Signs, streets, and storefronts: A history of architecture and graphics America's commercial corridors. *Technology and Culture, 54*(3). https://doi.org/10.1353/tech.2013.0089

Moon, S. (2005). *Militarized modernity and gendered citizenship in South Korea.* Duke University Press.

Seoul Metropolitan Government. (2008). *Signboard design guidebook.* ebook.seoul.go.kr/Viewer/HURSCPNTS15J

Venturi, R., Scott Brown, D., & Izenour, S. (1977). *Learning from Las Vegas: The forgotten symbolism of architectural form.* MIT Press.

Victoria and Albert Museum. (2023, May 15). https://www.vam.ac.uk/exhibitions/hallyu-the-korean-wave

Chapter 19
Public Life on Stage: Temporary Urbanism as a Platform for Cultural Expression

Guido Robazza

Abstract This chapter explores the intersections of temporary urbanism with the performative aspects of urban life, examining how ephemeral interventions act as platforms for cultural expression, engagement, and the revaluation of urban spaces. It argues that temporary urbanism, through its transient and strategic projects, not only addresses immediate urban challenges but also serves as a vibrant stage for the performance of local cultures and social interactions. By exploring the historic role of public space as a space for exhibiting and performing social and political life, the study discusses Urban Living Labs as a method, highlighting their role in staging cultural life and fostering a deeper connection between citizens and the urban landscape. At the heart of our exploration is the recognition of these interventions as both artistic and cultural performances that invite urban dwellers to participate in the co-creation of their environment. This paper underscores how temporary urbanism initiatives stimulate local culture and awareness, transforming passive spaces into active realms of cultural expression and community involvement. Through case studies, we illustrate how these temporary interventions not only challenge conventional urban planning approaches but also promote a culturally enriched, inclusive, and dynamic participatory urbanism. Furthermore, the study highlights the significance of these spatial practices in enhancing the value and perception of urban spaces, encouraging residents to engage more in depth with their city. By spotlighting the cultural dimensions of temporary urbanism, this paper contributes to an enriched understanding of how temporary urban practices can ignite significant cultural and spatial transformations, ultimately leading to a more engaged, aware, and vibrant urban community.

Keywords Temporary urbanism · Performative aspects · Ephemeral interventions · Cultural expression · Urban engagement · Revaluation of urban spaces · Transient projects · Local cultures · Social interactions · Public space · Urban Living Labs · Cultural life · Citizen participation · Co-creation · Active realms · Community involvement · Case studies · Conventional urban planning ·

G. Robazza (✉)
School of Architecture, Art and Design, University of Portsmouth, Portsmouth, UK
e-mail: guido.robazza@port.ac.uk

A. Melis and M. Pelle (eds.), *Design for Performative Arts Spaces*,
https://doi.org/10.1007/978-3-031-98215-6_19

Participatory urbanism · Spatial practices · Value enhancement · Urban awareness · Cultural dimensions · Spatial transformations · Vibrant urban community

Public Space as Contested Space

Public space has, almost by virtue of its existence, the innate potential to serve as exhibitory space. Its openness and the ways in which it is constructed, utilized, and populated inherently convey messages to the citizens who interact with it. In recent decades, a primary concern regarding public space has involved the manifestation and reinforcement of, as well as challenges to, social and economic inequalities (Buitelaar et al., 2016; Miraftab et al., 2015). Urban Living Labs (ULL) defined by the European Network of Living Labs (ENoLL[1]) as 'user-centred, open innovation ecosystems based on a systematic user co-creation approach in public–private–people partnerships, integrating research and innovation processes in real life communities and settings', represent innovative approaches to democratizing these spaces through bottom-up temporary urban placemaking actions. Such interventions aim to return spaces to the citizens and transform selected areas in ways that allow the city to be viewed from fresh perspectives and used in unexpected ways.

The historical evolution of public space serves as a common thread in the narrative of urban history. Public space has transitioned from the agoras of ancient Greece and the Roman forums to the boulevards of Paris designed by Haussmann. Spaces as Central Park in New York, the medieval squares used for commercial exchanges and the streets of the Baroque era symbolized changing political identities. These spaces have never been politically neutral; they have always played a part in social reproduction and the communication of ideas about the city and society. Even under medieval feudal monarchies, these spaces were where power was demonstrated and legitimized (Hénaff & Strong, 2001) (Fig. 19.1).

In the modern era, even in states with the most totalitarian attitudes, the importance of public space as venue for exhibition and display is evident, as seen in Red Square in the Soviet Union and Tiananmen Square in the People's Republic of China (Marron, 2016). While these spaces can be deliberately used by such regimes for displays of public conformity, military might, or patriotic enthusiasm, their significance extends beyond such orchestrated events. The notoriety of Tiananmen Square, alongside sites like Maidan in Ukraine, and Gezi Park in Turkey, illustrates that these locations can also serve as platforms for non-conformism rallying, dissidence, and resistance (Hatuka, 2018).

Even in democracies, conventional urban planning often excludes the common citizen from the creation process. These citizens are the primary user of these spaces, and their opinions and rights to shape the shared environment are often ignored (Rius-Ulldemolins & Klein, 2020). Long before the French Revolution, "the public" in the Western tradition referred to a literate and educated circle (Agacinski, 2001), and

[1] https://enoll.org/about-us/what-are-living-labs/.

Fig. 19.1 I Heart Pompey

even in recent Western history, the public sphere has been characterized by significant exclusions based on gender, class, and race (Fraser, 2014; Leddy-Owen et al., 2018). Furthermore, the privatization of spaces and a growing attitude towards prohibition and discouragement of their free use often legally or conventionally exclude citizens from appropriating these spaces (Lara-Hernandez et al., 2019; Robazza et al., 2024).

In the modern era, cities have been traditionally designed as "growth machines," primarily driven by profit-oriented forces (Fainstein, 2014; Sassen, 2005). This approach has often led to exclusive environments, biased distribution of resources towards lucrative interventions. As a result, degraded neighbourhoods considered economically insignificant are neglected. Likewise, public space with the potential to be vibrant and authentic expressions of citizens' desires has fallen into disuse. This is due to neoliberal urban development strategies that pursue economic success and international competition (Lord & Tewdwr-Jones, 2014; Sager, 2011). City streets and squares are progressively being segregated from the rest of the city to create "safe havens" for shoppers and businesses. While these spaces may physically resemble traditional public space, their functions and meanings have become significantly limited. Increasingly, public funds are used to subsidize the development of private spaces, potentially at the expense of more inclusive development (Hou, 2010).

That isn't to say that all public spaces have become sterile and monocultural. As Whyte (1980) observed in his seminal book on small spaces, people can and will gather in any public space where they are not actively prevented from doing so, often turning these areas into impromptu exhibition spaces for buskers, performers, or campaigners. Such appropriation can be defined as "temporary acts in which people use public space to carry out activities, individual or collective, different from the purposes for which the space was originally designed" (Fonseca Rodriguez, 2015). These activities may include commercial activities, social and leisure rituals such as children's games, and even community initiatives. In "The Ludic City," Stevens (2007) explores the alternative and playful uses of urban spaces and how these foster new experiences and social relationships. As Ricky Burdett suggests, today's cities are being rapidly remade, distancing themselves from the now obsolete ideological and spatial model of the 1933 Athens Charter (Burdett, 2018). In this process, "urban space" has increasingly become a territory for negotiating public and private interests, managed by a constellation of politicians, architects, real estate agents, banks, financial institutions, and landowners (Van Es et al., 2014).

Temporary Urbanism

In recent years we have witnessed a radical change of approach to public space with the growth of "temporary urbanism". This approach uses methods that "look to improve the urban experience through often temporary, low-cost, quick-to-use strategies" which are "spontaneous, participatory and community-driven" (Courage, 2013). Bishop and Williams (2012) identifies the financial crisis of 2007 as one of the main catalysts that poured numerous young people towards such interventions, including architects, artists, professionals, university researchers and activists. Their interventions have largely aimed to reactivate spaces in neglected urban areas, for example, by establishing temporary cinemas and agoras, which can then transform them into permanently active public space (for example Folly for a Flyover[2] or Autobarrios San Cristobal[3]), or by proposing a series of temporary hotel rooms in the city that citizens can inhabit (Hotel Shabby Shabby[4]). Other projects aim to raise awareness among residents of how public space in cities is conceived and on the need for more active and shared use. They promote pop-up/mobile community installations that engage people interested in their living environment (e.g. Union Press,[5] Street Interrupted[6]), or organize global events or artistic initiatives (e.g. Park(ing)[7]).

These interventions have not only been limited to the actions of outsiders and activists. The European Union has long supported the use of temporary urbanism as part of a project of urban regeneration policies. Initially it was promoted through the

[2] Folly for a Flyover, 2011, by Assemble. https://assemblestudio.co.uk/projects/folly-for-a-flyover.

[3] Autobarrios Sancristobal, 2014, by Basurama. http://basurama.org/proyecto/autobarrios-sancristobal/.

[4] Hotels Shabby Shabby, 2014, by Raumlabor Berlin. https://raumlabor.net/hotel-shabbyshabby/.

[5] Union Press, 2012, by PublicWorks. https://publicworksgroup.net/projects/union-press.1/.

[6] Street Interrupted, 2010, by MUF Architecture/Art. http://muf.co.uk/portfolio/street-interrupted-2010/.

[7] Park(ing) Day, sincere 2005, by Rebar. https://www.myparkingday.org/.

project URBAN CATALYST (Strategies for Temporary Uses—Potential for Development of Urban Residual Areas in European Metropolises[8]) in 2001, led by the TU Berlin, followed by the Dutch OLE project (Open Lab Ebbinge[9]) in 2008. The TUTUR (Temporary Use as a Tool of Urban Regeneration[10]) project, launched in 2013, was used to study how the temporary use of public space and informal practices can be integrated and become a tool of government for urban regeneration. A noteworthy ongoing project financed by the European Union on Temporary Urbanism is T-Factor,[11] that aim to boost novel approaches to urban regeneration, focussing on the key role that temporary uses can play in unleashing inclusive, sustainable and thriving urban areas.

The importance of temporary urban planning has equally been recognized by the United Nations with the Global Public Space Toolkit, published by UN-Habitat in 2015, which praised temporary interventions for their ability to: "give meaning and urban quality to spaces awaiting redevelopment quickly…with a strong community involvement" (Garau, 2015). In fact, temporary urbanism sits precisely in accordance with Sustainable Development Goal n. 11 and article 13 of 2017's New Urban Agenda of Habitat III to ensure that all cities are "participatory, promoting civic engagement and arousing a sense of belonging among all their inhabitants" (United Nations (Habitat III), 2017).

Placemaking

Placemaking has become a common term for identifying bottom-up temporary urbanism interventions, but it also describes the way 'all of us as human beings transform the places in which we find ourselves into places in which we live' (Schneekloth & Shibley, 1995, p. 1). Placemaking not only alters the physical environment but also the connections and relationships we build with others in our community (Schneekloth & Shibley, 1995) but it is transformative only insofar as power is devolved to communities, allowing them to direct all parts of the process (Arnstein, 1969) and as such, it relies on effective community participation (Kelkar & Spinelli, 2016; Sara et al., 2020). For this reason, co-design and co-creation, considered the most effective methods to delegate control, should be considered fundamental to the methodology of temporary urbanism.

[8] Urban Catalyst. http://www.urbancatalyst.net/index.php?lang=en.

[9] Open Lab Ebbinge. https://www.linkedin.com/company/open-lab-ebbinge-foundation-/?originalSubdomain=nl.

[10] TUTUR. https://urbact.eu/tutur-complete-overview.

[11] T-Factor. Participatory Futures. Co-creating cities with temporary urbanism. https://www.t-factor.eu/.

Co-creation, Co-design, and Co-production

The concepts of co-creation and co-design emerged from the concept of participatory design, a concept used in various disciplines since the 1970s. Although originating in northern Europe to improve industrial production (Bannon & Ehn, 2012; Huybrechts et al., 2017; Lenskjold et al., 2015) participatory practices were swiftly adapted to become an important part of design practices worldwide, and by 1980 also had become part of the process of designing places.

Co-design is a specific instance of co-creation; the latter includes any act of creativity that is shared by two or more people, the former entails collective creativity across the whole design process. In co-design users play a central role in informing, ideating, and conceptualising the design product. Non-expert users co-design with field experts often using design innovation tools to adapt the design and shift responsibilities (Manzini, 2015). As such some scholars see co-design methodology as being at the top of the citizens' participation scale (Edwards-Schachter et al., 2012, p. 682).

It is relevant to mention that the value of placemaking lies not only in the final outputs of the process but in the process of collective production itself. Placemaking transforms the community in more ways than by granting immediate liveability benefits (Markusen & Nicodemus, 2018); it produces social capital through the engagement of those directly involved (Matarasso, 1997). As Guinard (2018) remarks, public art is not just an aesthetic tool, but a method for generating knowledge based on life in the local community, one recent study finding that 'collaboration, connectivity, creating space and altruism can be potential mechanisms leading to a stronger sense of belonging, social cohesion and inclusion' (Madsen, 2019). If artistic activities are conducted in a group environment, they can facilitate new social interactions (Askins & Pain, 2011), especially in groups that are diverse, as artistic activities often involve non-verbal communication (Macpherson & Bleasdale, 2012).

Co-production involves a range of issues of belonging, citizenship and ownership (Zebracki & Palmer, 2018) and the 'citizen is the co-implementer' of change (Voorberg et al., 2014). Some scholars and practitioners in the field of urban innovation directly relate co-production to this production of social values, where resilience is the main outcome, rather than to the physical artefact or practices that result (Petrescu et al., 2016). In line with this view, community-engaged temporary urbanism should ultimately be interpreted as a practice that aims at creating social value, as the most important outcome.

Urban Living Labs

At the Portsmouth School of Architecture, Art and Design we established an Urban Living Lab to work with several end user groups in Portsmouth across a series of interventions that were part of the *Co-creation of temporary interventions in public space as a tactic to co-produce community resilience* project, aimed at developing a long-term strategy on several sites across the city (Robazza, 2020). These interventions were designed around the consolidated methodology of Urban Living Labs, defined as 'user-centred, open innovation ecosystems based on a systematic user co-creation approach in public–private–people partnerships, integrating research and innovation processes in real life communities and settings' (Bulkeley et al., 2019; *European Network of Living Labs*, n.d.; Puerari et al., 2018; Steen & Van Bueren, 2017). These projects encompass eight main interventions, along with a few unbuilt designs: "I Don't Roll" (2015); "The Secret Garden" (2016); "#IHeartPompey" (2016); "The Sound Garden" (2017); "The Edge Pavilion" (2018); "The Chatterbox - Multicultural City Pavilion" (2019); "The Architecture of the Woods" (2022); and the "ViaLibera!" Project (2024). These temporary interventions provided an opportunity to iteratively reflect on the developments and learnings from these projects, while simultaneously studying their impact and the degree of engagement with, and by, the community of participants.

The development of the co-creational pilots started from the Double Diamond innovation framework (Design Council, 2021; Hawryszkiewycz & Alqahtani, 2020; Tschimmel, 2012). This framework is structured around four key non-linear phases: discover, define, develop and deliver. The first diamond encourages people to understand the problems and define the challenges. The second diamond focuses on developing answers and test solutions. This framework guided our pilot, which was highly participatory, iterative and used design tools to visualise ideas. This design framework also suggests a set of design methods, or tools, that are partially derived from the service design processes. As Manzini (2015) notes, design tools are fundamental in facilitating the conversation between expert designers and non-expert designers. The social innovation process should establish a dialogue between the interested actors, which has to be supported by various design tools at different stages and with different aims. This approach is grounded on theories of design (Chasanidou et al., 2015; Razzouk & Shute, 2012; Simon, 2019) and design-enabled innovation (Concilio et al., 2019) and how these practices are transferred to the urban realm.

The process brought together many different agents in the city, encouraging the emergence of a dynamic system of relationships based on the common interests between key public organizations, such as academic institutions, city authorities and the third sector. Through them we saw first-hand how a public space could become an exhibitory realm for expressing sense of community pride, cultural values or even simply an expression of empowerment through having taken ownership and active use of the space.

Even without intervention, all public space both have the capacity to be used as, and to be, exhibition space; acting, in essence, as an exhibition, of values, of power dynamics, in and of themselves. Just as Hausmann's boulevards and avenues in Paris defined a closing off of public spaces from democratic control for the visual and political benefit of elites (Da Costa Meyer, 2022) by democratising public space by handing placemaking power to the users of these spaces, even if only temporarily, they become an active display of these democratising values and trends (Oureilidou,

2018). As such even a temporary intervention can both create and personify a statement of democratisation from which future social reproduction and placemaking can only further build.

References

Agacinski, S. (2001). Stages of democracy. *Public space and democracy* (pp. 129–143). University of Minnesota Press.

Arnstein, S. R. (1969). A ladder of citizen participation. *Journal of the American Planning Association, 35*(4), 216–224. https://doi.org/10.1080/01944366908977225

Askins, K., & Pain, R. (2011). Contact zones: Participation, materiality, and the messiness of interaction. *Environment and Planning d: Society and Space, 29*(5), 803–821. https://doi.org/10.1068/d11109

Bannon, L. J., & Ehn, P. (2012). Design: Design matters in participatory design. In *Routledge International handbook of participatory design* (pp. 37–63). Taylor and Francis. https://doi.org/10.4324/9780203108543

Bishop, P., & Williams, L. (2012). *The temporary city*. Routledge.

Buitelaar, E., Raspe, O., & Weterings, A. (2016). Urban inequality and justice. Creating conceptual order and providing a policy menu. *PBL Working Paper Series*.

Bulkeley, H., Marvin, S., Palgan, Y. V., McCormick, K., Breitfuss-Loidl, M., Mai, L., von Wirth, T., & Frantzeskaki, N. (2019). Urban living laboratories: Conducting the experimental city? *European Urban and Regional Studies, 26*(4), 317–335.

Burdett, R. (2018). Contemporary Urbanism. In *The Quito Papers and the new urban agenda* (pp. 140–149). Routledge.

Chasanidou, D., Gasparini, A. A., & Lee, E. (2015). Design thinking methods and tools for innovation. *Lecture Notes in Computer Science (Including Subseries Lecture Notes in Artificial Intelligence and Lecture Notes in Bioinformatics)* (Vol. 9186, pp. 12–23). https://doi.org/10.1007/978-3-319-20886-2_2

Concilio, G., Cullen, J., & Tosoni, I. (2019). Design enabled innovation in urban environments. In G. Concilio, J. Cullen, & I. Tosoni (Eds.), *Innovation capacity and the city. Springer Briefs in Applied Sciences and Technology* (pp. 85–101). Springer. https://doi.org/10.1007/978-3-030-00123-0_5

Courage, C. (2013). The global phenomenon of tactical urbanism as an indicator of new forms of citizenship. *Engage 32: Citizenship and Belonging*, *32*, 88–97. www.engage.org/journal

Da Costa Meyer, E. (2022). *Dividing Paris: Urban renewal and social inequality, 1852–1870*. Princeton University Press.

Design Council. (2021). *What is the framework for innovation? Design Council's evolved Double Diamond*. Design Council. https://www.designcouncil.org.uk/news-opinion/what-framework-innovation-design-councils-evolved-double-diamond

Edwards-Schachter, M. E., Matti, C. E., & Alcántara, E. (2012). Fostering quality of life through Social Innovation: A Living Lab methodology study case. *Review of Policy Research, 29*(6), 672–692. https://doi.org/10.1111/J.1541-1338.2012.00588.X

European Network of Living Labs. (n.d.). Retrieved August 30, 2021, from https://enoll.org/

Fainstein, S. S. (2014). The just city. *International Journal of Urban Sciences*, *18*(1). https://doi.org/10.1080/12265934.2013.834643

Fonseca Rodriguez, J. M. (2015). La importancia y la apropiación de los espacios públicos en las ciudades. *Revista de Tecnologia y Sociedad, 4*(7), 1–11.

Fraser, N. (2014). Rethinking the public sphere: A contribution to the critique of actually existing democracy. In *Between borders* (pp. 74–98). Routledge.

Garau, P. (2015). *Global public space toolkit: From global principles to local policies and practice.* www.unhabitat.org
Guinard, P. (2018). Redefining publics, artists, and urban spaces: The case of Made in Musina, South Africa. *City and Society, 30*(1).
Hatuka, T. (2018). *The design of protest: Choreographing political demonstrations in public space.* University of Texas Press.
Hawryszkiewycz, I., & Alqahtani, A. (2020). Integrating open innovation process with the Double Diamond design thinking model. In *European Conference on Knowledge Management.*
Hénaff, M., & Strong, T. B. (2001). The conditions of public space: Vision, speech, and theatricality. In M. Hénaff & T. B. Strong (Eds.), *Public space and democracy.* University of Minnesota Press.
Hou, J. (2010). Insurgent public space: Guerrilla urbanism and the remaking of contemporary cities. *Routledge.* https://doi.org/10.4324/9780203093009
Huybrechts, L., Benesch, H., & Geib, J. (2017). Co-design and the public realm. In *CoDesign* (Vol. 13, Issue 3, pp. 145–147). Taylor and Francis. https://doi.org/10.1080/15710882.2017.1355042
Kelkar, N. P., & Spinelli, G. (2016). Building social capital through creative placemaking. *Strategic Design Research Journal, 9*(2), 54–66. https://doi.org/10.4013/sdrj.2016.92.01
Lara-Hernandez, J. A., Melis, A., & Lehmann, S. (2019). Temporary appropriation of public space as an emergence assemblage for the future urban landscape: The case of Mexico City. *Future Cities and Environment, 5*(1), 1–22. https://doi.org/10.5334/FCE.53/METRICS/
Leddy-Owen, C., Robazza, G., & Scherer, L. (2018). Participatory design, temporary structures and the appropriation of urban space by marginalized youth: The problem of the Odd Triangle. *Journal of Urban Design, 23*(5), 780–797. https://doi.org/10.1080/13574809.2018.1458583
Lenskjold, T. U., Olander, S., & Halse, J. (2015). Minor design activism: Prompting change from within. *Design Issues, 31*(4). https://doi.org/10.1162/DESI_a_00352
Lord, A., & Tewdwr-Jones, M. (2014). Is Planning "under attack"? Chronicling the deregulation of urban and environmental planning in England. *European Planning Studies, 22*(2), 345–361. https://doi.org/10.1080/09654313.2012.741574
Macpherson, H., & Bleasdale, M. (2012). Journeys in ink: Re-presenting the spaces of inclusive arts practice. *Cultural Geographies, 19*(4), 523–534. https://doi.org/10.1177/1474474012442820
Madsen, W. (2019). Re-creating community spaces and practices: Perspectives from artists and funders of creative placemaking. *Journal of Applied Arts & Health, 10*(1), 25–40. https://doi.org/10.1386/jaah.10.1.25_1
Manzini, E. (2015). *Design, when everybody designs: An introduction to design for social innovation.* MIT Press.
Markusen, A., & Nicodemus, A. G. (2018). Creative placemaking. Reflections on a 21st-century American arts policy initiative. In C. Courage & A. McKeown (Eds.), *Creative placemaking. Research, theory and practice* (1st ed., Issue NOV-DEC, pp. 11–27). Routledge. https://doi.org/10.4324/9781315104607-2
Marron, C. (2016). *City squares: Eighteen writers on the spirit and significance of squares around the world.* HarperCollins.
Matarasso, F. (1997). *Use or ornament? The social impact of participation in the arts.* Comedia Publishing Group.
Miraftab, F., Wilson, D., & Salo, K. E. (2015). Introduction. In F. Miraftab, D. Wilson, & K. E. Salo (Eds.), *Cities and inequalities in a global and neoliberal world.* Routledge.
Oureilidou, E. (2018). Planning the cultural and social reactivation of urban open spaces in Greek metropoles of crisis. In *Defining landscape democracy* (pp. 189–199). Edward Elgar Publishing.
Petrescu, D., Petcou, C., & Baibarac, C. (2016). Co-producing commons-based resilience: Lessons from R-Urban. *Building Research and Information, 44*(7), 717–736. https://doi.org/10.1080/09613218.2016.1214891
Puerari, E., de Koning, J. I. J. C., von Wirth, T., Karré, P. M., Mulder, I. J., & Loorbach, D. A. (2018). Co-creation dynamics in Urban Living Labs. *Sustainability (Switzerland), 10*(6). https://doi.org/10.3390/su10061893

Razzouk, R., & Shute, V. (2012). What is design thinking and why is it important? *Review of Educational Research, 82*(3). https://doi.org/10.3102/0034654312457429

Rius-Ulldemolins, J., & Klein, R. (2020). From top-down urban planning to culturally sensitive planning? Urban renewal and artistic activism in a Neo-Bohemian District in Barcelona. *Journal of Urban Affairs*. https://doi.org/10.1080/07352166.2020.1811114

Robazza, G. (2020). Build art, build resilience. Co-creation of public art as a tactic to improve community resilience. *The Journal of Public Space, 5*(4). https://doi.org/10.32891/jps.v5i4.1388

Robazza, G., Priego-Hernández, J., Caputo, S., & Melis, A. (2024). Temporary urbanism as a catalyst for social resilience: Insights from an Urban Living Lab practice-based research. *Buildings, 14*(6), 1513. https://doi.org/10.3390/buildings14061513

Sager, T. (2011). Neo-liberal urban planning policies: A literature survey 1990–2010. *Progress in Planning, 76*(4), 147–199. https://doi.org/10.1016/J.PROGRESS.2011.09.001

Sara, R., Jones, M., & Rice, L. (2020). Austerity urbanism: connecting strategies and tactics for participatory placemaking. *CoDesign*, 1–17. https://doi.org/10.1080/15710882.2020.1761985

Sassen, S. (2005). The global city: Introducing a concept. *Brown Journal of World Affairs, 11*(2), 27–43.

Schneekloth, L. H., & Shibley, R. G. (1995). *Placemaking: The art and practice of building communities.*

Simon, H. A. (2019). *The sciences of the artificial* (3rd ed.). MIT Press.

Steen, K., & Van Bueren, E. (2017). *Urban Living Labs: A living lab way of working*. Delft University of Technology.

Stevens, Q. (2007). *The ludic city: Exploring the potential of public spaces*. Routledge.

Tschimmel, K. (2012). Design thinking as an effective toolkit for innovation. In: *Proceedings of the XXIII ISPIM Conference: Action for Innovation: Innovating from Experience.*

United Nations (Habitat III). (2017). *New urban agenda.* www.habitat3.org

Van Es, E., Harbusch, G., Maurer, B., Perez, M., Somer, K., & Weiss, D. (2014). *Atlas of the functional city, CIAM4 and comparative urban analysis*. Thoth Publishers & gta Verlag.

Voorberg, W. H., Bekkers, V. J. J. M., & Tummers, L. G. (2014). A systematic review of co-creation and co-production: Embarking on the social innovation journey. *Public Management Review, 17*(9), 1333–1357. https://doi.org/10.1080/14719037.2014.930505

Whyte, W. H. (1980). The social life of small urban spaces. In *New York: Municipal art society*. Project for Public Spaces.

Zebracki, M., & Palmer, J. M. (2018). Introduction to special issue: Urban public art: Geographies of co-production. *City and Society, 30*(1), 5–13. https://doi.org/10.1111/ciso.12152

Chapter 20
Public Art for a Better Aura in Urban Spaces

Mohamed Yazid Khemri

Abstract Interest in Public art has been increasingly rising in the past few decades, with special attention to its contribution to urban regeneration. A wide range of scholars claim that public art not only contributes to the aesthetic of urban spaces, but also to develop a sense of place, tackle social exclusion, and promote community resilience. Drawing on case studies in both the global North and South, this chapter looks at how public art can be used as a tool of regeneration of urban spaces. It will explore how public art can improve the streets of Algiers and Birmingham. With Birmingham city council announcing huge budget cuts to arts and culture by 60% in 2024 and 100% in 2025, it is worth investigating and highlighting how public art was used to creatively improve the inherited industrial urban spaces of the second largest city in the United Kingdom. Equally, the chapter will examine public art as being a product of community engagement and resilience. The chapter will rely on case study research as a method of investigation combined with participant observations; to study how public art helps regenerate urban spaces and increase their social use. The chapter will first synthesize the literature on public urban art, its societal impacts and contributions to urban regeneration. After that, the chapter will present reflection on observations and site visits conducted in both Algiers and Birmingham to understand how public art was used to regenerate the area.

Keywords Public art · Urban regeneration · Aesthetic contributions · Sense of place · Social exclusion · Community resilience · Case studies · Global north and south · Algiers · Birmingham · Budget cuts · Arts and culture · Industrial urban spaces · Community engagement · Participant observations · Social use · Literature synthesis · Societal impacts · Site visits · Regeneration strategies

M. Y. Khemri (✉)
Birmingham School of Architecture and Design, Birmingham City University, Birmingham, UK
e-mail: yazid.khemri@bcu.ac.uk

A. Melis and M. Pelle (eds.), *Design for Performative Arts Spaces*,
https://doi.org/10.1007/978-3-031-98215-6_20

Introduction

Historically, the expansion of cities has always been a response to political, economic, demographic, and socio-cultural needs and aspirations. As cities grow, they get exposed to wide range of challenges which do not only impact the quality of the built environment, but also how people interact with it.

The rapid expansion of cities throughout history has led to several challenges, such as unplanned and uncontrolled urbanization, often in response to post-war and demographic growth. This trend has been facilitated by modernism, which facilitated rapid urbanization, yet its deterministic design often resulted in unattractive and unwelcoming spaces. Additionally, cities faced the challenge of built environment degradation due to wars, abandonment, and lack of maintenance. All of the above have had significant repercussions on people's perception of space and the liveliness of urban areas. Moreover, lifeless spaces could also be the result of lack of safety (Khemri et al., 2020) privatization (Kohn, 2004), technologization and gender-insecurity (Gumpert et al., 1994).

The lack of liveliness and urban life can in turn have negative impacts on the vitality of urban spaces, can lead to social exclusion, and loss of sense of place and sense of community (Appadurai, 1996; Khemri & Melis 2021; Khemri et al., 2021). This issue has been discussed before by anthropologists like Augé (1995) who refers to lifeless spaces as "non-places", philosopher and sociologist Henri Lefebvre (1991) in his book about the social production of space. Also, many other scholars and planners advocated for urban life such as Jacobs (1961), Gehl (1987, 2010); Whyte (1980), Lynch (1984), De Certeau (1984), and recently Khemri and Melis (2021, 2022) through their work on El Houma.

At the end of 1960s and as a reaction to the Modern movement and its deterministic design, humanistic design surged to revive existing and forgotten good practices and suggest new design strategies that prioritise humans over cars and promote urban life and social interaction. Since then, several efforts have been made to improve existing urban spaces, while some are driven by governments; others are initiated by community members. These strategies were implemented at different scales and for different objectives. They all fall under the concept of Urban regeneration, which is considered a solution to enliven urban spaces and improve their look and atmosphere.

Urban Regeneration Though Public Urban Art

According to UN-Habitat (2021); urban regeneration consists of a series of interventions, that seek to transform urban areas into lively, safe, inclusive, diverse, resilient, and sustainable neighbourhoods, with the aim to extend those positive impacts to the wider city scale.

A successful urban regeneration approach is integrated, inclusive and holistic that addresses physical, environmental, and socio-economic aspects of the built environment. As urban regeneration does not only aspire to improve the physical environment but also the social aspect through using resilience, social cohesion, community participation to promote urban life, socio-economic prosperity and co-creating the city (UN-Habitat, 2021).

One of the (currents) tools of Tactical Urbanism is public urban art, with its various manifestations in urban spaces. Arts-based placemaking is a comprehensive approach to urban regeneration and community building with the aim to promote cultural diversity, local economy, liveliness and civic engagement. The benefits of public art extend to represent culture and character of the area, and fuels place value is fuelled by and used to represent culture (Project for Public Spaces, 2007). Public urban art actively groups street artists together to (re)create, (re)produce and engage in place making (Conklin, 2012).

In fact, art and the built environment have historically been combined and complementary to each other. Throughout history, architects and craftsmen used ingenuity and artistry to create buildings and places that are engaging and aesthetically pleasing. Beyond that, artwork I was exhibited in public spaces through sculptures, and used as ornamentation on buildings' facades. To reflect the culture of the area and tell stories of the community, as well as emanating charm, character and sense of place. In this respect, Sitte (1889, 1945), who had a remarkable influence in urban planning thinking and design, romanticised and urge planners and architects to design according to artistic principles and move away from rigid and deterministic design principles.

Public art can take different forms and shapes, from sculptures, murals, urban furniture to ornaments on buildings…Public art has been used as a tactical urbanism tool to regenerate urban spaces. Tactical urbanism is a term used to describe bottom-up approaches, low-cost, often temporary interventions to improve and activate urban spaces. Tactical urbanism can be initiated by different actors and stakeholders, including government bodies, non-profit organizations, businesses, community groups and individuals (Balarezo & Karimi, 2017). Tactical urbanism falls under the overarching concept of Placemaking, which refers to the collaborative process of reclaiming and revitalising urban spaces (Khemri et al., 2020).

It contributes to enlivening the area, and helps residents reclaim and redesign spaces that reflect their identity and socio-cultural way of life, allowing citizens to explore new forms of city making (Lydon & Garcia, 2015). It also empowers communities to take ownership of their urban environment, foster sense of belonging and improve social resilience (Robazza et al., 2024).

Public urban art has the ability to capture peoples' attention and encourages them to gather and talk as it stimulates discussions and reactions, whereas people would always have an opinion about the artwork, its message, meaning, and cultural value (Balarezo & Karimi, 2017).

One the other hand, some forms of urban art like graffiti can be seen as a crime, whereas in the United Kingdom graffiti is treated as an offence of criminal damage. Furthermore, people who see graffiti as vandalism; cite the broken window theory, which claims that the presence of misdemeanours and vandalism stimulates more serious crimes and urban decay because of visual disorder (Conklin, 2012; Sampson, 2009). Nonetheless, according to Balarezo and Karimi (2017) this theory has never been proven with quantifiable evidence and the perception of graffiti is changing and being revalorized.

Furthermore, public art as a tool for regeneration needs to be looked at from different angles, for instance, the presence of art and artists in neighbourhoods can indicate urban regeneration, community enhancing initiatives but also gentrification. Urban artists often appropriate de-valorised inner-urban residential zones, and derelict buildings and revalorise them through arts and other aesthetic and cultural contributions (Cameron & Coafee, 2005). Consequently, these revalorised and improved areas then attract people with higher economic capital, which can be considered as outsiders by the community, increase place and house value making it unaffordable to local community, all of which eventually lead to the disappearance of the character of the area (Cameron & Coafee, 2005) and causing people to relocate as they longer afford to live in the area (Khemri & Melis 2021; Khemri et al., 2021).

Observations Across Algiers and Birmingham

The methodology used to conduct this investigation is ethnographic and primarily qualitative. Starting with a literature review to understand the role of public art in urban regeneration. Then, visits and observations around neighbourhoods of Algiers and Birmingham where buildings and urban areas have been covered with public arts. Informal conversations with local people also took place to get their perception on the impacts of public urban art.

The visits conducted in Algiers took place in the Casbah which is a medieval Medina that was founded in 944, by the Berber Zirid Sanhaja dynasty, Bologhine Ibn Ziri, and developed by the Ottomans sixteenth–nineteenth century and the French nineteenth–twentieth century during their colonisations of Algeria (Khemri & Melis, 2022).

The casbah has always attracted visitors to explore its medieval architecture represented in the form of mosques, castles, neo-Moorish houses, and museums as well as its Algerian folklore. Throughout time, the medina started to degrade and decay, which in addition to its convoluted and narrow streets sent signs of lack of safety, which caused spaces to be lifeless. As a reaction, members of El Houma (Algerian form of community: see Khemri & Melis, 2022) resorted to public art to regenerate the area, improve its aesthetic and the atmosphere. This resulted in attracting tourists to the area, increasing liveliness, raising awareness about architectural heritage and folklore of the area, and boosting local economy.

Interestingly, the artwork used in the area differ in its form; from mural arts, to hanged paintings or setting up artisanal shops showcasing traditional handmade objects and artefacts representative of culture and identity of Algeria. The paintings were also illustrations of local architecture and monuments, and socio-cultural lifestyles (see Fig. 20.1). All of this strengthen the cultural identity and sense of place of the area.

As for Birmingham, the city council declared bankruptcy in 2023, which meant a revision of its strategy, targets, and priorities. Unfortunately, the city recently announced huge budget cuts to arts and culture by 60% in 2024 and 100% in 2025, despite having relied on public art to regenerate the area and represent its culture.

In Digbeth, which is an inherited industrial area in Birmingham that for years has had a reputation of rough and unsafe area due to its industrial, abandoned, and derelict buildings and urban spaces. It has been going through an urban regeneration process and public urban art was used throughout the area to create a better atmosphere as well as showcasing culture and identity of the area. The aim of regenerating Digbeth was to enliven the area, boost local economy, promote liveliness to increase natural surveillance and reduce crimes and attract people to live and work in the area.

Fig. 20.1 Public urban art used in the streets of the Casbah to regenerate the area, attract visitors and promote safety and sense of place. *Source* Author

Digbeth is now the city's premier creative and artists quarter, with a powerful cluster of creative and digital businesses, independent shops, and event venues. An artwork in the form of "The Green Man" statue serves as a reminder of nature and symbolises growth and rebirth.

The success of the TV show Peaky Blinders has also attracted hundreds of visitors wanting to track the history of this notorious gang who once lived in Digbeth and Small Heath. In the same respects, Mural arts of Peaky Blinders are now scattered throughout Digbeth and other parts of Birmingham, on commercial buildings and pubs with the aim of attracting visitors and tourists to boost the local economy.

Furthermore, in Birmingham city centre, the artworks represent the diversity of architecture and rich history the city has. Artworks can range from paintings, colourful pieces, to sculptures and statues. They tell stories of the different important events and people that influenced Birmingham throughout history. A case in point is "The Golden Boys" or "The Carpet Salesmen" sculpture in Broad Street, who represent Boulton, Watt and Mudoch; three pioneers of the industrial revolution. Another example is the sculpture of Thomas Attwood, in Chamberlain square, who was a significant political reformed and one of Birmingham's first two MPs. Other artworks represent people from the city like "A real Birmingham Family" sculpture which depicts single mothers and sisters Roma and Emma Jones and their sons, or the steel statue of Birmingham born comedian Tony Hancock. The "Iron Man" sculpture pays homage to the industrial revolution era and represents the skills developed in Birmingham and the Black Country. Mosaic was also used to represent figures like the first Irish Lord Mayor of Birmingham, Mike Nangle, Kennedy, and Martin Luther King Jr to inspire and bring hope to the heart of Digbeth (Crabbe, 2023).

Conclusion

Public urban art is sometimes used as a regeneration tool that represents political and socio-cultural aspirations of the city and its inhabitants. It contributes to bringing new life to decaying, derelict and abandoned urban spaces in the city and transforming them into vibrant, lively and energetic pocket spaces.

Public urban art in its various forms, aims to improve the atmosphere, the physical and aesthetic quality of the city, the social environment, and the economic well-being.

Public urban art can also promote adaptive reuse of abandoned buildings, as urban art can attract visitors which will trigger buildings to be converted into restaurants, cafes, galleries… furthermore, when public urban art is performed by local residents; sense of community and local identity are promoted, and residents develop a strong connection with their neighbourhoods increasing place attachment and sense of belonging.

Public art is not sustainable or resilient solution, but temporary as the aim of tactical urbanism. Public art can provide a better aura in public spaces, that are derelict, abandoned, and lifeless. Whereas regeneration can bring life back to the area and sends signs of growth, prosperity, sense of community and sense of place.

Nevertheless, it is also worth noting that urban regeneration is a more complex intervention that should not be summarised or limited to public art.

Even though urban public art is used as an urban regeneration tool, it is important to acknowledge that public art can't solve everything, and other hidden and more serious issues require more resilient, sustainable, and rigorous solutions. For instance, lifeless urban spaces can't only be solved through public art, but through a series of interventions and considerations of safety, walkability, facilities, urban furniture, legibility, comfort...

References

Appadurai, A. (1996): *Modernity at Large: Cultural dimensions of globalization. The public worlds series*. University of Minnesota Press.

Augé, M. (1995). *Non-places: Introduction to an anthropology of supermodernity* (J. Howe, Trans.). Verso.

Balarezo, M. L. G., & Karimi, K. (2017). Urban art and place. Spatial patterns of urban art and their contribution to urban regeneration. In *Polytechnic University of Valencia Congress, 24th ISUF 2017—City and Territory in the Globalization Age Conference Proceedings.*

Cameron, S., & Coaffee, J. O. N. (2005). Art, gentrification and regeneration—From artist as pioneer to public arts. *European Journal of Housing Policy, 5*(1), 39–58. https://doi.org/10.1080/14616710500055687

Conklin, T. R. (2012). *Street art, ideology, and public space*. Portland State University.

Crabbe, R. (2023). Public art in Birmingham. Public Art in Birmingham—Staying Cool at Rotunda.

De Certeau, M. (1984). *The practice of everyday life*. University of California Press.

Gehl, J. (1987). *Life between buildings: Using public space*. Van Nostrand Reinhold.

Gehl, J. (2010). *Cities for people*. Island Press.

Gumpert, G., & Drucker, S. J. (1994). Public space and urban life: Challenges in the communication landscape. *Journal of Communication, 44*(4), 169–177. https://doi.org/10.1111/j.1460-2466.1994.tb00706.x

Jacobs, J. (1961). *The death and life of great American cities*. Random House.

Khemri, M, Y., & Melis, A. (2021). Achieving community resilience through informal urban practices: The case of El Houma in Algiers. In Di Raimo, A., Lehmann, S., & Melis, A. (Eds.), *Informality through Sustainability; Urban informality now*. Routledge. https://www.taylorfrancis.com/chapters/edit/10.4324/9780429331701-23/achieving-community-resilience-informal-urban-practices-mohamed-yazid-khemri-alessandro-melis

Khemri, M. Y., & Melis, A. (2022). Algiers as a paradigmatic Mediterranean City: El Houma as a measure of urban life. *disP—The Planning Review, 58*(2), 16–39. https://doi.org/10.1080/02513625.2022.2123162

Khemri, M. Y., Caputo, S., & Melis, A. (2021). The drawbacks of a global concept of sustainable neighbourhoods in developing countries. In N. Mohareb, A. Versaci, Y. Mahgoub, S. Maruthaveeran, & F. Alberti (Eds.), *Cities' vocabularies: The influences and formations*. Springer. https://doi.org/10.1007/978-3-030-51961-2_16

Khemri, M. Y., Melis, A., & Caputo, S. (2020). Sustaining the liveliness of public spaces in El Houma through placemaking: The case of Algiers. *The Journal of Public Space*. https://www.journalpublicspace.org/index.php/jps/article/view/1254

Kohn, M. (2004). *Brave new neighborhoods: The privatization of public space*. Routledge.

Lefebvre, H. (1991). *The production of space*. Blackwell.

Sampson, R. J. (2009). Disparity and diversity in the contemporary city: Social (dis) order revisited. *The British Journal of Sociology, 60*(1), 1–31. https://doi.org/10.1111/j.1468-4446.2009.01211.x

Lydon, M., & Garcia, A. (2015). *Tactical urbanism. Short-term action for longterm change*. The Streets Plans Collaborative Inc.

Lynch, K. (1984). *Good city form*. The MIT Press.

Project for Public Spaces. (2007). What is placemaking? Available through https://www.pps.org/article/what-is-placemaking

Robazza, G., Priego-Hernández, J., Caputo, S., & Melis, A. (2024). Temporary urbanism as a catalyst for social resilience: Insights from an urban living lab practice-based research. *Buildings, 14*, 1513. https://doi.org/10.3390/buildings14061513

Sitte, C. (1889). *City planning according to artistic principles*. Phaidon.

Sitte, C. (1945). *The art of building*. Reinhold Publishing Corporation.

UN-Habitat. (2021). *Urban regeneration as a tool for inclusive and sustainable recovery*. Report on the Expert Group Meeting.

Whyte, W. H. (1980). *Social life in small urban spaces*. Conservation Foundation.

Chapter 21
Female-Themed Murals in the South Bronx as Expressions of Reclaimed Urban Spaces

Barbora Melis

Abstract This chapter explores the role of female-centered murals in reshaping urban narratives in the South Bronx. These murals embody performative urbanism, where art and social activism intersect to challenge traditional urban landscapes and amplify the voices of women within these spaces. Melis emphasizes that the murals transcend their visual appeal by representing resilience, empowerment, and the struggles of women within marginalized communities. Through vibrant expressions, these murals transform public spaces into arenas of social advocacy, addressing issues such as gender equality and safety. In particular, the South Bronx's historical and cultural context of economic hardship, community activism, and cultural creativity highlights the murals' role in advocating for social justice and community identity. Utilizing a methodology combining photography and qualitative research, this study documents how female-themed murals foster critical reflection on urban inclusivity, contributing to a renewed understanding of urban spaces that accommodate the experiences and identities of women and marginalized voices.

Keywords Female-themed murals · South Bronx · Urban narratives · Performative urbanism · Art and social activism · Traditional urban landscapes · Women's voices · Resilience · Empowerment · Marginalized communities · Vibrant expressions · Public spaces · Social advocacy · Gender equality · Safety · Economic hardship · Community activism · Cultural creativity · Social justice · Community identity · Photography methodology · Qualitative research · Urban inclusivity · Critical reflection · Renewed understanding · Experiences of women · Marginalized voices · Cultural context · Artistic transformations · Urban space redefinition · Social commentary · Identity representation · Intersection of art and activism · Public art impact · Visual culture · Community engagement · Historical context · Transformative art · Advocacy through art · Urban resilience

B. Melis (✉)
New York Institute of Technology, NY, USA
e-mail: bmelis@nyit.edu

A. Melis and M. Pelle (eds.), *Design for Performative Arts Spaces*,
https://doi.org/10.1007/978-3-031-98215-6_21

Introduction

In the dynamic landscape of urban environments, performative urbanism stands as a compelling force, reshaping how communities interact with and reclaim their public spaces. Defined as the intersection of artistic expression and social activism, performative urbanism reflects the evolving narratives of cities, where art becomes a powerful medium for dialogue, protest, and transformation. Nowhere is this more evident than in the South Bronx, a vibrant neighbourhood with a rich cultural tapestry woven from decades of socio-economic challenges and resilient community activism that inspires hope and resilience.

This contribution explores the phenomenon of female-themed murals in the South Bronx as an emotional expression of women reclaiming their spaces within the urban fabric. These murals, decorated with vibrant colours and compelling narratives, transcend aesthetics; they serve as visual testaments to women's identities through lived experiences, struggles, and triumphs in an often-overlooked urban context. Examining these murals, we explore the more profound questions of representation, identity, and empowerment within the public sphere.

These murals, beyond their artistic appeal, play a crucial role in challenging traditional urban narratives. They invite critical reflection on the pressing issues of gender equality and inclusion in urban planning and design. By doing so, they challenge us to reconsider urban spaces as physical environments and arenas where gender dynamics shape daily experiences, influencing safety, accessibility, and civic engagement. Through this exploration, we navigate the complexities of urban life through a lens that amplifies voices often marginalised in traditional urban narratives, fostering a sense of awareness and commitment to gender equality in urban planning.

This contribution aims to underscore the transformative power of art in reshaping urban landscapes. By examining the South Bronx as a case study, we uncover nuanced insights into how artistic interventions, such as these murals, can foster community resilience and advocate for social justice. These interventions have the potential to empower individuals and engage communities in the process of urban development, making them a powerful tool for change.

Historical and Cultural Context of the South Bronx

The South Bronx, located in New York City, is a densely populated neighbourhood with a rich cultural heritage and a history of resilience and hardship. It is home to a diverse population, including a significant proportion of Hispanic and African American residents. Their contributions have fostered a strong sense of community identity and a vibrant cultural scene. This diversity is vividly reflected in the area's murals, which serve as powerful expressions of collective memory, cultural pride, and social activism. Although globally renowned as the birthplace of hip-hop, the South Bronx remains America's poorest urban congressional district.

Originally a thriving industrial and residential area in the late nineteenth and early twentieth centuries, the neighbourhood underwent significant transformations throughout the twentieth century, shaped by waves of immigration, economic fluctuations, and urban planning policies. At the height of industrialisation, the South Bronx was home to bustling factories, residential brownstones, and vibrant immigrant communities. However, the mid-twentieth century witnessed a decline in industrial jobs and economic disinvestment, leading to widespread poverty, housing deterioration, and social unrest. The neighbourhood became synonymous with urban blight and neglect, exacerbated by factors such as redlining and discriminatory housing practices. Evolving from a historically low-profile and deprived area segregated from wealthy Manhattan (Hughes, 2021), the neighbourhood exemplifies today's shifting urban dynamics within New York City (NYU Furman Center, n.d.)

In recent decades, the South Bronx has experienced a resurgence fuelled by community-led revitalisation efforts, affordable housing initiatives, and cultural preservation projects. Organisations and local residents have worked tirelessly to restore historic buildings, promote economic development, and celebrate the neighbourhood's rich cultural heritage (Ciampa, 2018; Mokha, 2011). The legacy of resilience and creativity continues to shape the South Bronx as a dynamic and evolving urban landscape.

Amidst these challenges, the South Bronx emerged as a vibrant centre of cultural and artistic expression. In the 1970s and 1980s, the borough became synonymous with hip-hop music, graffiti art, and community activism. Artists and activists from diverse backgrounds used creative forms of expression to reclaim public spaces, challenge stereotypes, and advocate for social justice. Graffiti art, in particular, became a powerful medium for marginalised communities to assert their cultural identities and critique socio-political inequalities (Jonnes, 2022).

The murals of the South Bronx, often depicting strong female figures and themes of empowerment, stand as powerful symbols of resilience and advocacy. They provide a compelling backdrop for examining the complexities of inclusivity in urban public spaces, promoting gender equality and social justice, and offering deeper insights into the role of art in reclaiming urban spaces and empowering marginalised communities. The historical and cultural context of the South Bronx, with its rich legacy of overcoming adversity and embracing cultural diversity, underscores the transformative power of art in revitalising urban environments. By exploring the evolution of the South Bronx through this lens, we understand how art can serve as a catalyst for social change and community empowerment in urban settings.

Female-Themed Murals: Artistic Expression and Reclamation

New York City is famed for its vibrant street art scene with murals and graffiti that decorate its many open spaces, featuring iconic artists like Kobra, Banksy, and Keith Haring. Street art, often featuring large letters and vibrant colours in a style known as "tagging," has been an integral part of Bronx culture for decades. In NYC, it emerged in the 1970s, a time when the city grappled with bankruptcy and high crime rates. Growing up in this environment, kids from the Bronx and Brooklyn tagged subway cars and buildings, despite their actions' illegal and risky nature.

As the art form gained popularity, it also faced criticism as an act of vandalism, leading the city to ban subway graffiti in 1989. Despite this, artists persisted in using the streets as their canvas, and eventually, art collectors and curators acknowledged the value of their work. Today, street artists are often sponsored by business owners, community groups, and developers (Weaver, 2024). Due to their growing popularity, cities now use street art to develop their community's individuality (Adams, 2020).

Conrad (1995) states that murals could be considered the most inclusive form of art in the United States, representing the concerns of ordinary people, accessible to everyone, and bearing heartfelt aspirations or visions for the future. Further, community murals have an essential expressive and didactic role by clarifying, expanding, and strengthening community values. This art form not only teaches and provokes thought, posing profound questions, but can also be visually compelling. They foster community pride and commitment to justice, educating outsiders about the struggles faced by traditionally oppressed communities. Community murals, in particular, are a democratic art form with a significant capacity to enhance and broaden multicultural understanding (Adams, 2020).

The South Bronx appears as a canvas, with colourful and vibrant murals adorning the walls and making profound statements about reclaiming urban spaces. Often overlooked in mainstream urban art narratives, these murals are living reflections of the experiences, hopes and challenges faced by the people, including women, in the urban fabric. Women are celebrated as the subjects of powerful female-themed murals and as the talented artists creating these vibrant works of art.

Throughout the borough, there are many murals that focus on female themes. They depict a wide range of narratives, from historical figures and cultural icons to everyday women challenging stereotypes and asserting their presence. These murals add beauty to the urban landscape and challenge traditional ideas about who should be seen in public spaces. They tell stories of resilience, cultural pride, and strength in the face of adversity. The murals aim to take back public spaces from male-centric narratives and give a voice to historically marginalized groups, sparking conversation and addressing issues like gender inequality. By portraying women in various roles and contexts, the murals encourage people to rethink their perceptions of gender and women's contributions to city life.

While beyond the Bronx, the stories of the women depicted in murals and the local female graffiti artists remain relatively unknown, these artworks ultimately demonstrate the transformative power of art to rejuvenate neglected areas, empower communities, and reshape urban landscapes.

Methodology: Triangulation and Photography as a Research Tool

Understanding the impact and significance of female-themed murals in the South Bronx requires a methodological approach that blends rigorous research techniques with the expressive nature of art. To explore the transformative power of graffiti, the methodology of triangulation, and the role of photography as an essential tool in uncovering the deeper layers of meaning embedded within these urban artworks.

The research was primarily conducted through literature review, nonintrusive observation, photographing, and semi-structured interviews through photo-elicitation. This comprehensive approach aligns with the triangulation methodology in qualitative research, which employs multiple methods or data sources to understand a given phenomenon thoroughly. This approach enhances the validity of the study by cross-verifying data obtained from diverse sources, providing a robust analysis. The convergence of data from these methodologies ensures a more comprehensive and validated study, instilling confidence in the research's findings (Constable et al., n.d.).

In the context of studying murals, triangulation allows us to approach the subject from different perspectives, artistic, socio-cultural, and spatial, thereby providing a comprehensive understanding of their impact on urban environments and community dynamics.

Photography serves as a qualitative research method, engaging in the analytical exploration of the visual. According to Ball and Smith, the precise record of material reality carries significant power within photography, advocating for photographs to be the subjects of investigation. They argue that photographic material often remains under-analysed, typically serving as no more than illustrative devices. They stress the importance of using photographic material as a valuable source of data worthy of thorough analysis (1992, pp. 12, 14).

In content analysis, objectivity is a vital criterion for the adequacy of positivistic research. Photographs, being instantaneous representations of reality, act as mirrors of reality with memory (Ball & Smith, 1992, p. 16).

However, as Blinn and Harrist (1991, p. 176) point out that photography transcends its role as a mere recording device. It is evolving into a method with significant visual and ethnographic research potential. The scientific study of photographs for purposes beyond aesthetics or documentation is expanding and gaining recognition.

Photography plays a crucial role in documenting and preserving these ephemeral artworks, allowing them to reach a wider audience and sustain their impact over time. Norman (1991) emphasizes that photographs may not have an obvious connection with the signs and objects they signify. However, their symbolic nature illuminates their potential for analytical and interpretive discourse. By capturing these visual stories, photography ensures that the messages of empowerment, identity, and resistance conveyed through murals and graffiti continue to inspire and provoke dialogue long after the paint has faded.

Through photography, the historical documentation of murals and graffiti can be preserved (Gonzalez, 2024) and the stylistic choices, thematic motifs, and symbolic gestures embedded within can be analysed, uncovering subtle nuances in artistic expression and further enabling the trace of the evolution of murals as they interact with the dynamic urban landscape of the South Bronx.

Furthermore, photography helps engage stakeholders and community members by visually conveying the stories and messages depicted in these murals. By sharing photos with wider audiences, we hope to amplify the voices of the artists and community members involved in creating and maintaining these impactful urban artworks.

The significance of photography in observing social and political conditions in deprived or critical contexts, which has contributed significantly to overcoming cultural stigma, reaping benefits, and even safeguarding the most vulnerable individuals, seamlessly aligns with the use of photographs as a research tool. This approach can enhance inclusivity by capturing and presenting various perspectives, cultures, identities, and experiences.

By using triangulation and photography as complementary research tools, this study aims to emphasise the significant influence of murals with female themes in the South Bronx, providing a nuanced understanding of how art can act as a driver for social change and community empowerment in urban environments.

Examples of Graffiti and Mural Artwork in the South Bronx

The South Bronx features numerous murals and graffiti artworks embody resilience, empowerment, and social justice. These creations by diverse artists enhance the urban landscape and make powerful statements on gender equality and community identity. They play a crucial role in reclaiming public spaces for marginalised voices, transforming walls and facades into canvases for expression and advocacy, and fostering a sense of belonging and pride within the community. By examining these artworks, we gain deeper insights into how art can challenge societal norms, promote social justice, and empower individuals in urban environments.

The marginalization and oppression faced by black and Hispanic youth led them to form graffiti crews as a way to establish a sense of identity and belonging. Graffiti became a means of connecting with others' experiences. As one source stated, "The main purpose of graffiti is to have a positive impact… it provides individuals with a way to express their emotions in the Bronx, a sense of purpose in life. The graffiti and hip-hop culture extends beyond the final artworks, paying tribute to those who have passed away" (DeFazio, 2024).

In certain areas like Bruckner Boulevard in Mott Haven, restaurants and bars have emerged alongside the art venues. Organizations have emerged in the Bronx to nurture artists working in this medium, provide opportunities, and highlight the significance of street art for the community. One such organization is WallWorks New York in Port Morris, an art gallery that features the works of both emerging and established artists, blending downtown and uptown art styles (DeFazio, 2024).

The Bronx, renowned as the birthplace of hip-hop, has consistently been at the cultural forefront. The intertwined history of graffiti and music is evident, as the four foundational elements of hip-hop—graffiti, MCing, DJing, and breakdancing—were closely connected in their early development.

The marginalisation and oppression of Black and Hispanic youth fostered a sense of identity and belonging through the formation of graffiti crews. Among the most prolific graffiti artists from the South Bronx is Tats Cru, who has collaborated with several hip-hop legends. They began their journey by spraying trains as teenagers, later transitioning to painting on walls, and continue to be highly productive today. Their community murals play a crucial role in preserving the culture and history of the Bronx. According to the members of Tats Cru, these crews provide a sense of camaraderie, protection, and mutual support (DeFazio, 2024).

The Puerto Rican artist Bikismo created a mural of JLo [Jennifer Lopez] in celebration of memorable moments in her career and life and to honour the release of her documentary film "Halftime" and her commitment to social responsibility, accountability, and representation. He is renowned for his exceptional hyper realistic artworks created solely with spray paint and has created several pieces for prominent sports figures and singer (Leaden, 2022).

The art gallery WallWorks, founded by the renowned graffiti artist John Matos, also known as "Crash," and directed by his daughter Anna, aims to reflect and enhance the neighbourhood's cultural essence while honouring hip-hop. The gallery also provides a space for women artists to connect with the community and celebrate one another. This initiative merges and extends two previous projects: The Bronx Graffiti Art Gallery and Her Story. The First Annual Female Graffiti Series featured artists like Lady K Fever, Mel1, Rocky 184, Charmin 65, Miss Boombox, Gem13, and Neks. The mission of Her Story is to create a supportive environment for female writers to share their narratives and showcase their skills (Stavsky, 2016, 2023).

This focus on female representation and empowerment through art leads to a broader discussion on women's perspectives on the city and gender equality.

Women's Perspectives on the City and Gender Equality

Despite recent progress, architecture, urban planning, and design have been historically known as male-dominated professions, resulting in a one-sided perspective on the urban realm. This disparity can contribute to creating gender inequalities that significantly impact women, girls and sexual and gender minorities, especially in comparison to cisgender, heterosexual men. Further, the intersection of gender with other characteristics of vulnerability, such as poverty, disability, race or ethnicity, and age, introduces additional inequalities that compound this impact. All these groups experience the city through a series of physical, social, economic, and symbolic obstacles that profoundly influence their daily lives in gendered ways, which may not be as visible to men who do not encounter these urban challenges.

Many planning and design practitioners still lack a comprehensive understanding and acceptance of these challenges, not fully grasping how conditions in the built environment, coupled with a lack of diversity in those shaping it, contribute to and sustain gender inequity. Incorporating a gender perspective into urban planning and design is essential for creating inclusive and equitable cities. Women, in particular, face challenges while navigating cities in multifaceted roles as caregivers, workers, and community members. Their experiences inform how they perceive and use cities, leading to unique challenges compared to many men. This situation creates a disparity between the physical layout of urban spaces and how individuals, especially from a gendered perspective, experience and navigate those environments, especially when it comes to aspects of safety.

The concept and application of "woman" as an analytical category in the present study may pose challenges, as it tends to obscure differences originating from characteristics other than gender, such as race, ability, class, and age. Therefore, in the present study, the terms and concepts of "her," "woman," and "female" refer to all subjectivities who understand themselves as such, independent of anatomy.

In the South Bronx, it's important to consider women's viewpoints when thinking about urban spaces. This is crucial for dealing with safety, accessibility, and inclusivity issues in public areas. Several factors, such as social, economic, and cultural elements, shape how women experience urban environments. Given the historical and socio-economic gaps and ongoing efforts to renew the South Bronx, women have to navigate public spaces while dealing with varying levels of safety concerns and access to resources.

Examining these experiences through qualitative research methods, such as interviews, reveals the nuanced ways in which women interact with and perceive their urban surroundings. These insights provide valuable perspectives on the lived realities of women in the South Bronx and inform strategies for enhancing gender-responsive urban planning and design.

Urban design plays a pivotal role in shaping gender equality within cities. From the layout of streets and public transportation systems to the provision of public amenities and recreational spaces, urban planners can create environments that promote equitable access and participation for all residents, regardless of gender.

Addressing gender inequities in urban settings requires a concerted effort to challenge existing norms and advocate for inclusive practices. In the South Bronx and beyond, community-led initiatives and grassroots movements play a pivotal role in reclaiming public spaces and advocating for policies that promote gender equality.

These efforts highlight the need to confront and address the systemic differences that persist in urban environments.

Conclusion

In urban environments, performative urbanism is reshaping how communities interact with and reclaim their public spaces. This art-meets-activism phenomenon is vividly illustrated in the South Bronx. Female-themed murals there serve as powerful expressions of women reclaiming their urban spaces, transcending aesthetics to become visual testaments to women's identities, struggles, and triumphs.

These murals challenge traditional urban narratives and prompt critical reflection on gender equality and inclusion in urban planning. They encourage us to view urban spaces not just as physical environments, but as arenas where gender dynamics profoundly shape daily experiences, influencing safety, accessibility, and civic engagement.

The South Bronx's historical and cultural context underscores the transformative power of art in revitalizing urban environments. The neighbourhood's murals, often depicting strong female figures, serve as symbols of advocacy and resilience, promoting gender equality and offering more profound insights into the role of art in reclaiming urban spaces and empowering marginalized communities.

The intersection of art and urban development is further highlighted by the role of photography as a research tool. Photography captures and preserves the ephemeral nature of murals, ensuring that the messages conveyed through these artworks continue to inspire and provoke dialogue long after the painting has faded.

Examining female-themed murals in the South Bronx emphasizes art's significant influence in promoting social change and community empowerment in urban settings. This exploration serves as a reminder of the enduring power of art to challenge societal norms, advocate for justice, and foster inclusive and equitable urban environments.

References

Adams, M. (2020). *Understanding murals influence on the community*. Queens University of Charlotte.

Ball, M. S., & Smith, G. W. H. (1992). *Analyzing visual data. Qualitative Research Methods Series* (Vol. 24). Sage. https://doi.org/10.4135/9781412983402

Blinn, L., & Harrist, A. W. (1991). Combining native instant photography and photo-elicitation. *Visual Anthropology, 4*(2), 175–192. https://doi.org/10.1080/08949468.1991.9966559

Ciampa, F. (2018). Participatory design approach for the Bronx waterfront. *Proceedings of The 6th International Virtual Conference on Advanced Scientific Results, 6*, 202–206. https://doi.org/10.18638/scieconf.2018.6.1.499

Conrad, D. (1995). Community murals as democratic art and education. *The Journal of Aesthetic Education, 29*(1), 98–102. https://www.jstor.org/stable/3333522

Constable, R., Cowell, M., Zornek Crawford, S., Golden, D., Hartvigsen, J., Morgan, K., Mudgett, A., Parrish, K., Thomas, L., Turner, Erika Yolanda Thompson, R., & Palmquist, M. (n.d.). *Ethnography, observational research, and narrative inquiry. Types of qualitative observational research*. The WAC Clearinghouse. Colorado State University. Retrieved February 7, 2024, from https://wac.colostate.edu/repository/writing/guides/

DeFazio, C. (2024). Exploring the history & the Bronx with Tats Cru. *UP Magazine*. https://upmag.com/up6-tats-cru/

Gonzalez, D. (2024, August 17). A South Bronx graffiti walkabout. *The New York Times*. https://archive.nytimes.com/lens.blogs.nytimes.com/2015/08/17/a-south-bronx-graffiti-walkabout/

Hughes, C. J. (2021). In the Bronx, Mott Haven Suddenly Gets a Skyline. *The New York Times, Okt. 17*, 12. https://www.nytimes.com/2021/10/15/realestate/mott-haven-bronx-new-construction.html

Jonnes, J. (2022). *South Bronx rising*. Fordham University Press.

Leaden, C. (2022). *There is now a massive JLo mural in the Bronx*. Secret NYC. https://secretnyc.co/jlo-mural-nyc/

Mokha, K. (2011). Mott Haven tries to rekindle momentum. *Wall Street Journal*, 1–5. https://online.wsj.com/article/SB10001424052748704099704576289213167219704.html

Norman Jr., W. R. (1991). Photography as a research tool. *Visual Anthropology*, *4*(2). https://doi.org/10.1080/08949468.1991.9966560

NYU Furman Center. (n.d.). *Neighborhood profiles: Mott Haven/Melrose BX01*. Retrieved October 23, 2021, from https://furmancenter.org/neighborhoods/view/mott-haven-melrose

Stavsky, L. (2016). Her story, the first annual female graffiti series, launches uptown with Lady K Fever, Mel1, Rocky 184, Charmin 65, Miss Boombox, Gem13 and Neks. *Street Art NYC*. https://streetartnyc.org/blog/2016/05/16/her-story-the-first-annual-female-graffiti-series-launches-uptown-with-lady-k-fever-a-tribe-called-mel-rocky-184-charmin-65-leeza-boombox-gem13-and-neku/

Stavsky, L. (2023). *Lady K Fever on the NYC graffiti women's festival at Bronxlandia with: Lovenotes, Nasa One, Chare, Flô, Kstar, Alice Mizrachi, Mrs, Claw Money, Lexi Bella, Mikki Mu and More*. Street Art NYC. https://streetartnyc.org/blog/2023/10/26/lady-k-fever-on-the-nyc-graffiti-womens-festival-at-bronxlandia-with-lovenotes-nasa-one-chare-flo-kstar-alice-mizrachi-mrs-claw-money-lexi-bella-mikki-mu-and-more/

Weaver, S. (2024). The top spots to see graffiti and street art in NYC. *Time Out*. https://www.timeout.com/newyork/art/street-art-top-ten-spots-to-see-street-art-and-graffiti-in-nyc

Chapter 22
Visual Culture and Palestinian Identity in Protracted Displacement

Dana Hamdan

Abstract This chapter examines how visual and spatial practices shape identity within Al Wehdat refugee camp, focusing on the dual themes of resistance and assimilation among Palestinian refugees in Jordan. Through iconographic murals and material interventions in domestic architecture, residents convey their collective memory, cultural heritage, and future aspirations. Semi-structured interviews with long-term refugees reveal community-led initiatives aimed at beautifying public spaces and reasserting symbolic ownership of place. By integrating elements of both Palestinian and Jordanian heritage, these artistic practices embody a complex interplay of nostalgia, resilience, and adaptation, transforming the camp into a culturally charged landscape of belonging. Ultimately, this study underscores the vital role of creative expression in navigating the refugee experience adds nuance to debates on visual culture, place-making, and identity within protracted displacement.

Keywords Refugee Artistic expression · Al Wehdat refugee camp · Architectural adaptations · Collective memory · Community-led initiatives · Palestinian and Jordanian heritage

Introduction

Al Wehdat refugee camp, established in 1955 in Amman, Jordan, symbolises resilience and adaptation within the Palestinian diaspora. Initially home to 5000 refugees from the 1948 Arab–Israeli war, it now hosts about 60,000 long-term residents and new generations born into displacement (UNRWA, 2022a). Like other Palestinian camps, Al Wehdat serves as a repository of collective memory and identity. Despite challenges such as overcrowding and resource scarcity, its inhabitants persist under difficult economic and political conditions.

D. Hamdan (✉)
School of Architecture, Technology and Engineering, University of Brighton, Brighton, UK
e-mail: d.hamdan@brighton.ac.uk

A. Melis and M. Pelle (eds.), *Design for Performative Arts Spaces*,
https://doi.org/10.1007/978-3-031-98215-6_22

As one of 58 recognised Palestinian refugee camps in the Levant and one of 10 camps in Jordan (UNRWA, 2022b), Al Wehdat exists in a relatively stable political context, having seen no major internal conflict since the 1971 Jordanian civil war (Fruchter-Ronen, 2008; Kelberer, 2017). The Jordanian government has sought to assimilate 1948 refugees, primarily through citizenship, fostering a unique identity that blends belonging in Jordanian society with the preservation of Palestinian heritage and the right of return (Bastaki, 2020). However, those displaced in 1967, primarily from Gaza, remain non-citizens and face challenges to their assimilation. Their reluctance to assimilate, fearing the loss of their right of return to Palestine, combined with Jordanian policies that limit welfare responsibility for them, has left these refugees under the care of the United Nations Relief and Works Agency (UNRWA) and reduced their social, political and economic assimilation within Jordan (Feldman, 2012). This article will only present examples from the refugee camps established post 1948 dysphoria, not the 1967 one.

This article focuses specifically on Al Wehdat camp, home to generations of post-1948 refugees who, despite some political assimilation, live under the dual management of the Amman Municipality and the United Nations Relief and Works Agency for Palestine Refugees (UNRWA). Economic limitations, intergenerational trauma, and strong social ties keep them connected to the camp. The complex identity of these residents reflects the ongoing tension between assimilation and the desire for return, which is often expressed through public artistic endeavours (Alnsour & Meaton, 2014; Hamdan et al., 2024).

The primary objective of this article is to examine the relationship between artistic expression and identity among long-term Palestinian refugees in Jordan, using Al Wehdat as a case study. This research contributes to cultural preservation, ensuring that the stories and identities of Palestinian refugees are not lost. By highlighting Al Wehdat's hybrid identity, the study validates the resilience of its inhabitants and provides insights into how cultural preservation fosters social cohesion and resilience. Ultimately, this research enriches academic discourse on refugee studies and urban sociology, offering a nuanced understanding of how displaced communities navigate their identities through creative practices.

The Role of Artistic Expression and Shelter Adaptation as Social Capital in Refugee Camps

Hanna et al. (2022) argues that displacement does not equate to social and cultural dispossession for refugee communities. Refugees can maintain their cultural identity, practices, and traditions even while living in camps or settlements. Domestic refugee architecture refers to the built environments that these communities create, incorporating traditional elements from their homelands. This architecture involves adapting shelters to meet community needs, often challenging standard humanitarian designs.

It reflects the socio-spatial relations and cultural practices of refugees, shaped by necessity and their lived experiences (Hanna et al., 2022).

Domestic refugee architecture not only helps communities maintain cultural identity but also fosters social capital through relationships and cooperation. AlWaer et al. (2023) discuss how social connections in refugee communities can promote economic development, highlighting that investing in social capital—such as building trust and collaboration and improvement of public spaces—can improve business opportunities and overall well-being. This investment is crucial for addressing poverty, inequality, and job creation in camps (AlWaer et al., 2023; Yaghi et al., 2024). In Al Wehdat, artistic and architectural practices are vital for enhancing community support and resilience. These creative expressions contribute to cultural preservation and resistance while strengthening social ties and cohesion within the community (Hamdan et al., 2024).

The Role of Artistic Expression as a Form of Resistance in the West Bank

On the other hand, artistic expression has long served as a potent form of resistance in Palestinian territories. The graffiti on the Israeli Segregation Wall exemplifies this phenomenon, transforming a symbol of oppression into a canvas for resistance. Hasan and Bleibleh (2023) describe the wall as a 'traumascape' disrupted by the construction, which has been reimagined into a 'resistancescape' through the power of graffiti. This graffiti serves as a medium for Palestinians to express their collective memory, national identity, and resistance against the Israeli occupation. It commemorates national symbols and figures, fosters imagination and hope for a liberated future, and creates a space for urban dialogues and agency. Such everyday resistance empowers Palestinians to claim their right to the city and demand spatial justice.

Research Methods

This essay forms part of a broader urban research project that documents long-term Palestinian refugee camps. The overarching study employs qualitative research methods grounded in ethnographic approaches, with fieldwork conducted over two years in Al Wehdat refugee camp. Data collection methods included participant observation, in-depth interviews, and visual ethnography, enabling a comprehensive understanding of the lived experiences of the camp's inhabitants and the spatial characteristics of their environment.

Observations and Discussion of Artistic and Architectural Resilience Within Al Wehdat Refugee Camp

Fieldwork in Al Wehdat refugee camp revealed the dynamic relationship between architecture and creative practice, showcasing a wealth of innovative artistic expressions despite significant spatial limitations and resource scarcity. Urbanised refugee camps like Al Wehdat often face challenges such as deteriorating construction materials, narrow pathways from households encroaching on public spaces—a phenomenon Maqusi (2021) refers to as "spatial violations"—and the incremental vertical growth of buildings to accommodate successive generations. Amidst these difficulties, residents have made notable efforts to personalise and beautify their living environments. This creativity is prominently displayed in the murals that adorn the camp's streets, which reflect the community's resilience and ingenuity. These murals serve not only as aesthetic enhancements but also as powerful symbols of collective memory and longing for the right of return. They capture elements of both Palestinian heritage and Jordanian influences, illustrating a dynamic interplay of cultural preservation and adaptation. This article focuses specifically on the artistic expressions within the camp's public outdoor environment, leaving the exploration of heritage mobility and adaptability in the refugee homes' interior layouts for further research.

The facades of housing units in Al Wehdat are undergoing a transformation through beautification efforts. Some units are clad with limestone, a material commonly found in Jordan and the Levant region in general. To mimic the appearance of basalt stone, residents have painted some of the limestone blocks black. This alternating pattern of dark and light stone is a hallmark of the Ablaq architectural style, reflecting a blend of local materials and traditional design elements from their cultural heritage. This Mamluk architectural technique is characterised by alternating rows of light-coloured stones interspersed with dark natural stones, creating a striking visual contrast. By painting certain stones darker to replicate this traditional aesthetic, residents preserve a significant aspect of their cultural heritage while enhancing the visual appeal of the housing structures (see Fig. 22.1).

Additionally, walking through the streets of Al Wehdat camp reveals vibrant murals adorning many housing units. The murals prominently feature Palestinian landmarks and symbols, evoking nostalgia for their homeland even after over 70 years of displacement. Some murals also incorporate Jordanian landmarks, illustrating the refugee's navigation of their identity between displacement and peaceful integration with their host. Figures 22.2, 22.3, 22.4 and 22.5 illustrates such examples.

In semi-structured interviews with long-term refugees born in the camp, many participants highlighted community-led initiatives aimed at beautifying public spaces within the limited environment. One participant noted, "We have many young volunteers, such as the 'Pioneers' (Al Mutamaizoon), who have revitalised several neighbourhoods by planting gardens and painting the walls. The improvements made in each upgraded neighbourhood have been maintained by the residents to this day".

The slingshot is a powerful symbol of Palestinian resistance, particularly during the first Intifada in the west bank and Gaza (1987–1993) (Alım, 2020). It embodied

Fig. 22.1 [top]: Housing units clad in limestone in Al Wehdat refugee camp, with their residents painting sections of the limestone blocks in black to imitate the appearance of basalt stone. [bottom]: Housing units clad in limestone in Al Wehdat refugee camp, with their residents painting sections of the limestone blocks in black to imitate the appearance of basalt stone

Fig. 22.2 A member of the Al Mutamaizoon team paints a mural on a housing unit in the camp, featuring the phrase "**We will return**" alongside a depiction of a Palestinian town. Photo is provided by Rania Alkouz. This phrase can be seen on many walls in Al Wehdat camp, emphasising the enduring hope and determination of the residents to reclaim their homeland and preserve their cultural identity despite decades of displacement

the David versus Goliath narrative of the underdog confronting a seemingly overwhelming force. Used predominantly by young Palestinians, the slingshot represented their determination to resist occupation despite limited resources, highlighting the courage and resilience of those standing up against oppression. This simple tool became an icon of empowerment, capturing the spirit of defiance and the ongoing struggle for justice and self-determination in the face of adversity. The presence of the slingshot drawings in the Palestinian camps outside of Palestine showcases the enduring legacy of this symbol as a means of expressing resistance and identity. In refugee camps like Al Wehdat, slingshot imagery serves not only as a reminder of past struggles but also as a rallying cry for future generations to continue the fight for their rights. Even though residents of Al Wehdat were not physically present in the territories during the Intifada, the spirit of resistance resonates deeply within the camp, creating a shared sense of solidarity with those who were.

The dynamics between permanence and temporariness in Al Wehdat camp are central to understanding its evolving built environment and social fabric. While officially designated as a "camp," Al Wehdat has long surpassed the temporary conditions that the term implies. Since its establishment in 1955, residents have incrementally

Fig. 22.3 Depicts traditional Palestinian plant patterns with the word "Palestine" inscribed on one of the residential homes. Additionally, it features an illustration of a woman dressed in traditional Palestinian attire, balancing a water jug on her head—reflecting a common task performed by women in rural areas of Palestine in the past

Fig. 22.4 [left] Depicts Jordanian flag and the Roman columns in Jerash, a prominent landmark in Jordan painted on a housing unit in Al Wehdat camp. It symbolises the residents' harmony with their host country, further illustrating the integration of Jordanian and Palestinian cultural symbols within the community; [right] Jordanian flag, which can also be observed in several areas in the camp

Fig. 22.5 Handala, the slingshot, and the Palestinian flag are frequently depicted on the walls of Al Wehdat camp. These powerful images collectively represent the struggle for Palestinian rights, embodying themes of defiance and hope. Created by political cartoonist Naji al-Ali in 1969, Handala is a significant national symbol and embodiment of the Palestinian people diaspora. In Arabic, Handala translates to "medicinal bitter desert fruit." The character is depicted as being ten years old and remains frozen at this age, reflecting the same age that Naji Al-Ali was when he was expelled from his village in 1948 (Haugbolle, 2013)

transformed the space through self-built extensions, infrastructural improvements, and artistic interventions, embedding a sense of permanence despite the absence of formal land ownership. The dual governance of the camp—under both UNRWA and the Greater Amman Municipality—reinforces this ambiguity, as it remains administratively distinct from the surrounding city while functionally operating as a dense urban neighbourhood. This tension between the temporary and the permanent is further reflected in the way residents navigate their identity: while some adaptations mirror urban integration, others deliberately maintain symbols of displacement, resisting full assimilation to preserve the right of return. Artistic and architectural expressions within the camp become mechanisms for negotiating this liminal status, transforming its physical and symbolic landscape beyond the restrictive framework of a "temporary" refugee space.

Overall, the murals in Al Wehdat camp, an example of a Palestinian camp outside the Palestinian territories, blend Palestinian heritage and Jordanian landmarks, reflecting a hybrid identity formed by displacement and assimilation. In the West Bank, however, murals primarily serve as expressions of resistance against Israeli occupation, transforming oppressive spaces into powerful symbols of defiance, as was discussed by Hasan and Bleibleh (2023). The murals and street art throughout Al Wehdat camp turn the otherwise utilitarian and often dilapidated structures into vibrant expressions of cultural pride and historical narratives. These artistic interventions serve multiple purposes: they beautify the camp, display a sense of hope and pride among residents, and create a visual record of their collective memory and identity cantered around displacement and the right of return. By embedding layers of meaning and emotional resonance into the physical environment, the murals affirm the camps position not just a place of refuge but a repository of cultural heritage and resilience. The visual appeal of these artworks also helps to counteract the harshness of the refugee camp environment, providing a sense of dignity and normalcy to its inhabitants. Beyond their cultural and symbolic significance, these artistic and architectural practices also play a crucial role in space-making—transforming the camp's-built environment into an active site of identity negotiation and community cohesion. The camp's-built environment reflects a tension between temporariness and permanence, as self-built structures and community-led modifications challenge the notion of the refugee camp as a transient space. While official narratives often frame integration as stability, residents of Al Wehdat continue to maintain strong connections to their Palestinian heritage. They artistic expression in the camp displays a unique refugee identity that balances integration within Jordanian society with the preservation of Palestinian heritage and the assertion of the right of return. By modifying their surroundings, residents assert control over their lived space, challenging imposed spatial constraints and reclaiming ownership of their environment despite their lack of legal tenure. These creative interventions demonstrate how informal practices contribute to place-making, reconfiguring the refugee camp from a space of marginalisation into one of belonging and continuity. This paper highlights the value of artistic and architectural adaptations in shaping refugee identity. Future studies could examine how similar artistic and architectural strategies are employed in other displaced communities worldwide, providing comparative insights into the intersection of space, identity, and resistance in protracted refugee situations.

References

Alım, E. (2020). The art of resistance in the Palestinian struggle against Israel. *Türkiye Ortadoğu Çalışmaları Dergisi*, *7*(1), 45–80. https://doi.org/10.26513/tocd.635076

Alnsour, J., & Meaton, J. (2014). Housing conditions in Palestinian refugee camps, Jordan. *Cities, 36*, 65–73. https://doi.org/10.1016/j.cities.2013.10.002

AlWaer, H., Sibley, M., & Cooper, I. (2023). Design and use of space in refugee camps: A case study of a contested terrain. *Proceedings of the Institution of Civil Engineers-Urban Design and Planning, 176*(3), 123–141. https://doi.org/10.1680/jurdp.22.00068

Bastaki, J. (2020). The meanings of citizenship between resettlement and return: The case of displaced Palestinians. *Citizenship Studies, 24*(2), 154–175. https://doi.org/10.1080/13621025.2020.1720605

Feldman, I. (2012). The humanitarian condition: Palestinian refugees and the politics of living. *Humanity: An International Journal of Human Rights, Humanitarianism, and Development*, *3*(2), 155–172. Retrieved from https://www.humanityjournal.org/wp-content/uploads/2014/06/3.2-The-Humanitarian-Condition.pdf

Fruchter-Ronen, I. (2008). Black September: The 1970–71 events and their impact on the formation of Jordanian national identity. *Civil Wars, 10*(3), 244–260. https://doi.org/10.1080/13698240802167991

Hamdan, D. M., El Faouri, B., Teba, T., & Yaghi, A. (2024). Evaluating 'pocket' public spaces in Al Wehdat refugee camp, Jordan: challenges and opportunities. *Urban Research & Practice*, 1–26. https://doi.org/10.1080/17535069.2024.2372488

Hanna, D. M., Buys, L., & Kumarasuriyar, A. (2022). Domestic refugee architecture in Jordan: A socio-spatial analysis of chaotic camps. *The Journal of Architecture, 27*(1), 44–70. https://doi.org/10.1080/13602365.2022.2062034

Hasan, D., & Bleibleh, S. (2023). The everyday art of resistance: Interpreting "resistancescapes" against urban violence in Palestine. *Political Geography, 101*, Article 102833. https://doi.org/10.1016/j.polgeo.2023.102833

Haugbolle, S. (2013). Naji al-Ali and the iconography of Arab secularism. *Visual Culture in the Modern Middle East: Rhetoric of the Image*, 231–258.

Kelberer, V. (2017). Negotiating crisis: International aid and refugee policy in Jordan. *Middle East Policy, 24*(4). Retrieved from https://www.bu.edu/pardeeschool/files/2017/12/Kelberer_MEP-Final-002.pdf

Maqusi, S. (2021). Acts of spatial violation: The politics of space-making inside the Palestinian refugee camp. *ARENA Journal of Architectural Research, 6*(1). https://doi.org/10.5334/ajar.324

UNRWA. (2022a). *Palestine refugees.* https://www.unrwa.org/palestine-refugees

UNRWA. (2022b). *Where we work.* https://www.unrwa.org/where-we-work/jordan#:~:text=More%20than%202%20million%20registered,refugee%20camps%20throughout%20the%20country

Yaghi, A., Abuhassan, L., Hamdan, D., Aalwneh, L., Khushamn, L., Tarawneh, S., & Goussous, J. (2024). Typologies of informal adaptations of communal spaces: A comparative study between different socio-economic spatial settings in Amman. *Urban Research & Practice*, 1–30. https://doi.org/10.1080/17535069.2024.2383991

Part IV
Future Perspective and Conclusion

Chapter 23
Generative AI and the Collective Canvas: Redefining Artistic Authorship and Performance

Alessandro Melis and Marco Pelle

Abstract This chapter examines the revolutionary role of generative AI in reshaping performative art. This final chapter, rather than summarizing the book's contents, returns to the foundational question of art categorization, challenging rigid classifications and emphasizing their limitations, especially during global crises. Generative AI, presented as a transformative force, disrupts traditional artistic frameworks, facilitating new blends of creativity and collective approaches to art. Generative AI's capabilities extend beyond mere automation to collaborative artistry, blending algorithmic computation with human creativity. By using generative adversarial networks (GANs) and variational autoencoders (VAEs), AI systems generate novel content across media, challenging notions of singular authorship and expanding the collective nature of creation. This technology redefines art spaces, as seen in digital and virtual installations where AI's role extends to real-time, audience-responsive art, introducing the concept of art as a live, interactive process rather than a static object. The chapter's conclusion incorporates case studies demonstrating this integrated approach, illustrating how generative AI supports both democratization and professional artistry. It proposes guidelines for designing performative spaces with AI, emphasizing interdisciplinary collaboration, adaptive design, and ecological awareness. The chapter also explores exaptation—the process of repurposing existing structures for new functions—to guide a new taxonomy of creativity in AI-integrated art. By advocating for AI's role as an augmentation of human ingenuity, this chapter foresees a future where generative AI transforms performative arts, fostering a synergy between human and machine creativity that advances both artistic and societal progress.

Keywords Generative AI · Collective canvas · Artistic authorship · Performative art · Art categorization · Rigid classifications · Global crises · Transformative

---

A. Melis (✉)
School of Architecture and Design, New York Institute of Technology, New York, USA
e-mail: amelis@nyit.edu

M. Pelle
Department of Social and Political Sciences, Bocconi University, Milano, Italy
e-mail: marco.pelle@unibocconi.it

A. Melis and M. Pelle (eds.), *Design for Performative Arts Spaces*,
https://doi.org/10.1007/978-3-031-98215-6_23

force · Traditional artistic frameworks · Collaborative artistry · Algorithmic computation · Human creativity · Generative adversarial networks (GANs) · Variational autoencoders (VAEs) · Novel content generation · Singular authorship · Collective creation · Digital installations · Virtual installations · Audience-responsive art · Interactive process · Static objects · Case studies · Democratization of art · Professional artistry · Designing performative spaces · Interdisciplinary collaboration · Adaptive design · Ecological awareness · Exaptation · New taxonomy of creativity · AI-integrated art · Augmentation of human ingenuity · Synergy between human and machine · Artistic progress · Societal progress · Innovative art spaces · Creative collaboration · Future of performative arts · Transformational impact

Instead of a traditional conclusion summarizing the contents of this book, we have chosen to revisit the initial discussion about the conventionalities of categories in art. This exploration into the origins of creativity emphasizes the philosophical implications of reification and taxonomies, which we deem risky in times of global crises. Rather than reinforcing conventional categorizations, our aim is to challenge and expand them.

Additionally, we have decided to focus our conclusion on the emergence of Artificial Intelligence (AI), particularly Generative AI, as a transformative force in performative art. This technology has the potential to further disrupt established categories and foster a new era of blended creativity and holistic art perspectives.

To illustrate these ideas, we have included final case studies showcasing the potential of this integrated approach. These examples provide a glimpse into what this new paradigm of performative art could look like. We then use these insights to outline a concise guide, which serves as a foundational handbook for performing arts. This guide is intended to be shared in pedagogical and educational settings, as well as to support scholarly research and professional practice in the fields of performing arts and multidisciplinary studies.

Digital and Virtual Performance Spaces

The rise of digital technology has revolutionized performance art, introducing virtual spaces that transcend traditional architectural constraints. This exploration highlights the impact of digital and virtual reality technologies on performance spaces, focusing on their transformative effects on the performer-audience dynamic and the new immersive experiences they offer (Manovich, 2001).

Generative AI represents a significant shift in the arts, merging technology with creativity in unprecedented ways. By fostering a collaborative relationship between human creativity and algorithmic computation, generative AI expands the scope of what is considered art. This technology challenges traditional notions of authorship and creativity while reviving collective and performative art forms, prompting a re-evaluation of the roles of the artist, audience, and medium in the creative process.

Generative AI encompasses algorithms and systems capable of producing new, original content based on learned patterns and data. Unlike traditional AI, which focuses on understanding and interpreting data, generative AI draws inspiration from this data to create novel outputs, ranging from text and images to music and beyond. This creative process is driven by machine learning models, particularly generative adversarial networks (GANs) and variational autoencoders (VAEs), which iteratively refine their outputs through a feedback loop, minimizing the difference between the generated work and human-created work (Elgammal et al., 2017; Liapis et al., 2018).

Historically, art has been perceived as the expression of an individual's vision, skill, and creativity. However, generative AI disrupts this notion by introducing a collaborative dynamic between human artists and AI algorithms. In this symbiotic relationship, AI acts as a co-creator, contributing its computational creativity to the artistic process. This collaboration democratizes art creation, enabling individuals without formal art training to participate in artistic endeavors. Additionally, generative AI projects often involve community input or interaction, further emphasizing their collective nature. Thus, generative AI can be viewed as a contemporary return to communal art forms, where the creation process and the resulting artwork are shared among a group rather than attributed to a single artist.

The rise of generative AI in art prompts complex questions regarding authorship and originality. Traditionally, the value of art has been closely tied to the artist's unique vision and craftsmanship. However, when AI generates art, it challenges the concept of the artist as the sole author, suggesting a shift towards collective authorship. This perspective aligns with historical movements that embraced communal artistic practices, such as Dadaism and Surrealism, which valued collective creation and the role of chance in art (Bostrom, 2014; Russell & Norvig, 2016). By leveraging generative AI, modern creators revisit these themes, emphasizing process and collaboration over individual ego.

Moreover, generative AI can be interpreted as a form of performative art. The process of creating art with AI involves a dynamic interplay where the artist sets parameters and guides the algorithm, which generates outputs that can be unpredictable and unique. This interaction between human intention and algorithmic randomness resembles a performance, where the artistic process is as significant as the final product. Live generative AI installations, where algorithms create art in real-time in response to environmental inputs or audience interactions, further enhance this performative aspect, making the creation process an integral part of the art itself (Crawford, 2020).

As we continue to explore generative AI, it becomes evident that this technology will play a crucial role in shaping the future of artistic expression. By blurring the lines between creator and creation, and between art and performance, generative AI is redefining the boundaries of the art world and ushering in a new era of creative possibilities.

The Evolution and Ethical Challenges of Artificial Intelligence

The pursuit of Artificial Intelligence (AI) aims to replicate and enhance human cognitive abilities through technology. The intellectual foundation of AI traces back to pioneers like Alan Turing, whose work established the theoretical underpinnings of computing and intelligence evaluation, and John McCarthy, who coined the term "artificial intelligence" at the Dartmouth Conference in 1956. This era marked the formalization of AI as a field of study, initially focused on rule-based systems designed to mimic human problem-solving and decision-making. Early AI relied on explicit instructions, limiting its adaptability in unpredictable situations (Russell & Norvig, 2016).

A transformative shift occurred in the late twentieth century with the advent of machine learning, moving from rigid, rule-based architectures to flexible, learning-based models. Machine learning involves creating algorithms that learn from and adapt to new data, allowing systems to process extensive datasets, identify patterns, and make autonomous decisions. The quality and representation of data became crucial, as they directly influence the effectiveness of machine learning algorithms (Bostrom, 2014; Crawford, 2020).

Once trained, AI models find applications in diverse fields such as healthcare diagnostics, financial analytics, autonomous vehicle navigation, and content creation in the entertainment industry. These models continuously evolve, leveraging ongoing feedback to enhance functionality and drive efficiency and innovation across various sectors.

However, the widespread adoption of AI technologies raises significant ethical concerns. Privacy issues are paramount, given AI's reliance on personal data. Additionally, biases in AI systems, stemming from skewed or incomplete datasets, can lead to inequitable decisions. The potential for job automation and worker displacement further fuels debates on AI's socio-economic impact. These challenges highlight the need for ethical guidelines in AI research and application, emphasizing transparency, accountability, and fairness (Crawford, 2020; Yan & Li, 2020).

The journey from basic rule-based algorithms to advanced machine learning models reflects not just technological progress but also a deeper understanding of the interplay between AI and human society. As AI integrates into daily life, the dialogue around its ethical, social, and economic implications becomes increasingly vital. Ensuring that AI development respects human values and promotes equity is essential (Russell, 2019).

As AI capabilities advance, the notion of artificial general intelligence (AGI) emerges. AGI aims to perform a wide range of tasks with human-like cognitive functions, opening extraordinary possibilities but also raising ethical and philosophical questions about the limits of machine intelligence and the potential for AI consciousness.

Another significant area is augmented intelligence, where AI systems enhance human intelligence rather than replace it. This collaborative approach aims to amplify human analytical and creative capabilities, envisioning a future where AI acts as an ally to human potential.

The global competition for AI leadership presents geopolitical implications, underscoring the need for international cooperation and regulatory measures to ensure AI advancements benefit humanity and do not exacerbate global disparities.

Fostering a comprehensive understanding of AI's impact requires interdisciplinary research spanning technology, ethics, social sciences, and public policy. Open, inclusive dialogues and partnerships are crucial to leveraging AI's benefits while addressing its challenges. This approach can pave the way for a future where AI harmonizes with ethical standards and human values, advancing societal progress and well-being.

Ethical Implications of AI in Society

The rapid evolution of artificial intelligence (AI) demands ethical vigilance to navigate its complexities and prevent potential disruptions to the environment and society. The ethical concerns surrounding AI often stem from our difficulty in fully grasping its intricate nature and far-reaching implications, particularly in relation to human creativity and evolution (Bostrom, 2014).

In 1996, Steven Jay Gould highlighted humanity's inherent propensity towards self-destruction, a characteristic that predates modern technological advancements. This suggests that the primary hazard of AI is not necessarily its potential to surpass human capabilities, but rather its capacity to amplify our existing tendencies and behaviors significantly (Gould, 1996).

Thomas Huxley's analogy of the chessboard, from his 1868 work "A Liberal Education and Where to Find It," serves as a metaphor for nature—and by extension, AI—as an impartial yet challenging opponent governed by natural laws. This allegory, despite criticism from modern evolutionary biologists, reflects a once-popular view that underscores the flawed belief in human supremacy over nature. This perspective has influenced our urban and technological landscapes and continues to shape the ethical discourse on AI, particularly regarding creativity and authorship (Huxley, 2010).

Concerns about AI's impact on creativity and authorship arise from fears of losing creative control, rooted in outdated dichotomies rather than an understanding of the creative process. Creativity, an evolutionary trait that has driven human innovation, is not diminished by AI; rather, it can be expanded and enhanced. AI's role in the creative process challenges us to rethink traditional notions of authorship and creativity, advocating for a broader, more inclusive view of intelligence and creative expression (Harari, 2015).

As we explore the ethical dimensions of AI's integration into society, it is crucial to adopt a nuanced perspective on intelligence, creativity, and their relationship with technology. Recognizing the limitations of existing paradigms and AI's potential as a catalyst for innovation, we must strive for an adaptive, reflective approach to AI ethics. This approach encourages us to transcend apprehensions and embrace the expansive possibilities that AI offers for growth and creative exploration (Wilson, 1984).

Ethical concerns about AI's products being aesthetically superior to human creations overlook the subjective nature of aesthetic judgment. Aesthetic value is not absolute but holds significance only for those who judge it. Therefore, fearing that humanity might be surpassed by machines in producing aesthetically pleasing outputs is paradoxical. The same applies to functional value: surpassing human standards in functionality raises the question of "for whom?" and "to what end?" AI might evolve to address needs incomprehensible or different from human ones, yet this should not inherently be a cause for concern.

AI, as an extension of human and ecological systems, seeks ecological balance much like any other entity. Understanding AI in this context alleviates fears of antagonism and highlights the importance of maintaining equilibrium. The risk of unbalancing this system could lead to mutual destruction, emphasizing the need for harmonious coexistence (Wilson, 1984).

Furthermore, the vast array of variables in the ethical discourse on AI suggests that we should be equally, if not more, concerned about other existential threats, such as viral mutations, bacterial resistance, or self-destructive human behaviors driven by fanaticism or weaponization. Addressing these broader threats requires a holistic approach that considers AI as one component of a complex, interconnected ecosystem (Harari, 2015).

In summary, the ethical implications of AI necessitate a comprehensive understanding of its potential to enhance human creativity and innovation while maintaining ecological and societal balance. By embracing a reflective and adaptive ethical framework, we can harness AI's benefits while mitigating its risks, ensuring that technology serves to advance human well-being and societal progress.

The Concept of Reification: A Cross-Disciplinary Examination

Reification, the process of treating abstract concepts as if they were tangible, real objects, invites an exploration across various critical texts in philosophy, sociology, and psychology. This notion, derived from the Latin word "res," meaning "thing," profoundly influences our understanding of societal structures and individual perceptions and interactions.

Philosophically, and particularly within Marxist theory, the concept of reification is deeply rooted in Karl Marx's *Capital* (1867). Marx introduces the idea of commodity fetishism, a specific type of reification prevalent in capitalist societies, where relationships between people assume the form of relationships between commodities. This concept is crucial for understanding how human labor and social interactions are masked by commodification, leading to a distorted perception of social relations (Marx, 1867).

Building on Marx's foundation, Georg Lukács, in *History and Class Consciousness: Studies in Marxist Dialectics* (1971), explores the broader impacts of reification on consciousness and societal organization under capitalism. Lukács delves into how human relations are objectified and alienated within capitalist structures, offering profound insights into the systemic dehumanization and mechanization of social interactions (Lukács, 1971).

In sociology, Peter L. Berger and Thomas Luckmann's *The Social Construction of Reality: A Treatise in the Sociology of Knowledge* (1966) examines how societal constructs come to be regarded as real. Although their work does not explicitly focus on reification, it critically analyzes the processes through which social constructs are universally accepted as tangible, resonating with the essence of reification (Berger & Luckmann, 1966).

Michel Foucault's pivotal works, *Discipline and Punish: The Birth of the Prison* (1977) and *The History of Sexuality, Volume 1: An Introduction* (1978), provide invaluable insights into the reification of power relations and knowledge into concrete societal practices and institutions. Foucault's analysis of how disciplinary mechanisms and discourses solidify power dynamics and norms within society, although not explicitly labeled as reification, aligns with the critical examination of how abstract concepts are materialized and accepted as immutable components of the social order (Foucault, 1977, 1978).

These foundational works collectively describe the complex nature of reification, offering a framework for understanding its role in capitalist society, the establishment and perpetuation of social norms, the dynamics of power relations, and the relationship between knowledge and societal structures. Reification emerges not merely as a theoretical concept but as a critical tool for dissecting how abstract notions are transformed into perceived concrete realities, deeply influencing both societal structures and individual consciousness.

The ethical implications of AI's societal integration necessitate an adaptive, reflective approach. Recognizing AI's potential to enhance human creativity and innovation, it is essential to maintain vigilance over its impact on societal balance. Concerns about AI's products being aesthetically superior to human creations overlook the subjective nature of aesthetic judgment. The value of aesthetics is not absolute but holds significance only for those who judge it. Therefore, fearing that humanity might be surpassed by machines in producing aesthetically pleasing outputs is paradoxical (Crawford, 2020).

Additionally, the concept of AI surpassing human functionality raises questions about the relevance and application of human-set standards. AI might evolve to address needs that differ from human ones, yet this should not inherently be a cause for concern. AI, as an extension of human and ecological systems, seeks ecological balance much like any other entity. Understanding AI in this context alleviates fears of antagonism and highlights the importance of maintaining equilibrium.

Thus, fostering a comprehensive understanding of AI's impact requires interdisciplinary research spanning technology, ethics, social sciences, and public policy.

Open, inclusive dialogues and partnerships are crucial to leveraging AI's benefits while addressing its challenges. This approach can pave the way for a future where AI harmonizes with ethical standards and human values, advancing societal progress and well-being.

The Impact of Reification on Taxonomies, AI, and the Performing Arts

Exploring taxonomy through the lens of reification provides insights into how we organize and understand knowledge, particularly when considering the philosophies of Michel Foucault. Taxonomy, a system of classification that arranges concepts, objects, or information into categories based on specific criteria, is crucial for navigating and comprehending complex structures. While traditionally used in biology to categorize life forms, taxonomy extends across various domains to structure information systematically.

Reification plays a significant role in taxonomic discussions by transforming human-devised classifications into seemingly natural and fixed divisions. This process can obscure the subjective and constructed nature of these categories, falsely presenting them as inherent properties of the classified objects rather than reflections of human judgment and societal conventions.

Michel Foucault, through his investigations into the archaeology of knowledge and analysis of discourses shaping epistemes (the knowledge frameworks of specific epochs), critically explores the power and implications of taxonomies in human sciences. In his seminal work, *The Order of Things* (Foucault, 1966), Foucault examines how different historical periods are defined by distinct epistemes that legitimize certain taxonomies and knowledge divisions, thereby shaping our perception of reality.

He challenges the reified nature of taxonomic categories, arguing that they are not universal truths but are shaped by historical, cultural contexts, and power relations. He posits that taxonomies work as mechanisms through which knowledge is organized, controlled, and disseminated, cautioning against their seeming neutrality. According to Foucault, these classifications define norms, distinguishing "normal" from "abnormal" and "licit" from "illicit," thus embedding exclusionary practices and power dynamics into the fabric of knowledge and society.

In the broader discourse on artificial intelligence (AI), the critique of binary oppositions and recognition of our evolutionary limitations highlight the complexities of categorization and the potential pitfalls of uncritically applying reified taxonomies. This perspective challenges the notion of a dichotomy between humanity and an emergent "artificial" intelligence as adversaries, suggesting instead that our technological creations are extensions of our evolutionary journey, reflecting our capabilities and limitations.

This discussion on AI and reification bridges seamlessly with the exploration of AI in the performing arts. In the realm of digital and virtual performance spaces, AI's role in transforming the performer-audience dynamic and creating new immersive experiences can be seen as an extension of the reification process. Just as taxonomies classify and structure knowledge, AI categorizes and analyzes vast datasets to generate creative outputs, challenging traditional notions of creativity and authorship. This collaboration between human creativity and AI algorithms broadens the scope of what is considered art and prompts a reevaluation of the roles of the artist, audience, and medium in the creative process.

Furthermore, the ethical implications of AI's integration into society necessitate an adaptive, reflective approach. Recognizing AI's potential to enhance human creativity and innovation, it is essential to maintain vigilance over its impact on societal balance. Concerns about AI's products being aesthetically superior to human creations overlook the subjective nature of aesthetic judgment. The value of aesthetics is not absolute but holds significance only for those who judge it. Therefore, fearing that humanity might be surpassed by machines in producing aesthetically pleasing outputs is paradoxical.

Additionally, the concept of AI surpassing human functionality raises questions about the relevance and application of human-set standards. AI might evolve to address needs that differ from human ones, yet this should not inherently be a cause for concern. AI, as an extension of human and ecological systems, seeks ecological balance much like any other entity. Understanding AI in this context alleviates fears of antagonism and highlights the importance of maintaining equilibrium.

Addressing the challenges posed by anthropocentric views, especially in the context of AI, requires a profound shift in understanding our place within the natural world. Recognizing the relationships defining the web of life and the dependence of human survival on the vitality of the biosphere is essential. By adopting a more holistic and inclusive approach to classification and technology development, we can foster technologies that promote sustainable development, enhance human welfare, and ensure the preservation of the planet for future generations, aligning technological advancement with ethical and ecological considerations.

Thus, exploring taxonomy and reification through the lens of AI and performing arts not only enhances our understanding of these concepts but also highlights the need for a reflective and adaptive approach to AI ethics. This perspective ensures that technological advancements serve to advance human well-being and societal progress while respecting and enhancing our interconnectedness with the natural world.

An Extended Taxonomy of Creativity

Within the discourse of this document, the apprehension surrounding artificial intelligence (AI) transcends the fear of it eluding human oversight to become a renegade force. More critically, the concern is rooted in the potential of AI to serve as an amplification of human imperfections, notably through the imposition of our rigid frameworks upon the fabric of reality. This shift in perspective moves the dialogue away from a scenario where AI gains autonomy and acts against its creators, focusing instead on how AI might magnify the inherent flaws and biases in our perception and interaction with the world around us. Adhering too stringently to the taxonomies we've constructed, without acknowledging the fluid and interconnected nature of the natural world and technological advancements, positions AI to potentially accentuate and expand upon these constraints, fostering developments that might be detrimental to humanity. Consequently, this discussion aims to delineate a concept of creativity facilitated by AI, advocating for a broader taxonomy that surmounts the prevailing binary categorizations. This endeavor draws inspiration from the evolutionary concept of exaptation, which illustrates how characteristics can evolve to serve purposes distinct from their initial functions.

Exaptation, a principle derived from evolutionary biology, delineates the phenomenon whereby traits gain new functions unrelated to their original evolutionary purposes. When this concept is applied to the realm of AI, it suggests a paradigm where AI's function transcends the simplistic binary of acting merely as a tool or a threat. Instead, AI emerges as a collaborator in the creative process, endowed with the capacity to contribute to human creativity and innovation in manners that surpass existing categorical limitations. This evolved taxonomy not only recognizes AI's potential to recontextualize its capabilities in the face of emerging challenges and opportunities but also promotes a refined understanding of AI as both a reflection and an enhancement of human intellect through a cooperative relationship (Gould & Vrba, 1982).

Embracing the concept of exaptation enables the cultivation of an environment in which AI serves as a catalyst for dismantling the entrenched categorizations that currently limit our cognitive processes. This approach heralds a shift towards a more dynamic and adaptable interaction between humans and technology, spurring the pursuit of novel avenues in creativity and problem-solving. The incorporation of AI into a broader and more adaptable taxonomy stands to augment our capacity for innovation, equipping us to tackle complex issues in more comprehensive and creative ways.

Therefore, recasting the narrative on Generative AI from a dichotomy of control and antagonism to one of mutual collaboration and shared evolution paves the way for fully leveraging the capabilities of artificial intelligence. By broadening our taxonomic frameworks to encompass more cohesive and interconnected categories, and by gleaning insights from the heuristic value of exaptation, we are better positioned to navigate the multifaceted risks and prospects AI presents. This strategy not only curtails the likelihood of adverse, self-inflicted outcomes but also unlocks new possibilities for creativity and advancement, heralding a future where AI and human ingenuity converge in the pursuit of progress and innovation.

The Transformative Impact of Generative AI on Art, Architecture, and Performative Arts

Generative AI is a catalyst for transformative change in art and architecture, leveraging cutting-edge technologies like Generative Adversarial Networks (GANs) and Variational Autoencoders (VAEs). These innovations reshape creative possibilities, offering novel methodologies for artistic and design endeavors while prompting a critical examination of traditional notions of creativity and ownership. Generative AI is not just a tool but a paradigm shift, altering how creative output is conceptualized and produced (Elgammal et al., 2017; Liapis et al., 2018).

In architecture and design, generative AI employs algorithms to envision diverse design solutions tailored to meet criteria related to efficiency, structural soundness, and optimal use of space (Akbarzadeh et al., 2018). By simulating environmental interactions like airflow and thermal dynamics, generative AI can contribute, for instance, to developing buildings that are both sustainable and effective (Kolarevic, 2019). Additionally, AI's ability to analyze data allows for the incorporation of user preferences and environmental considerations into the design process, enhancing the functionality and relevance of architectural projects (Veloso et al., 2020). This technological progress fosters a collaborative ethos where machine learning becomes integral to the creative process, offering innovative solutions and foresight into potential design challenges, thereby amplifying the collective creative potential of architects and designers (Ciftcioglu et al., 2019).

This evolution invites reflection on the essence of creativity and the significance of human contribution in creating outstanding visual arts and designs. Generative AI blurs the lines traditionally separating artificial from natural and human creativity from AI, opening new horizons for creative expression. It cultivates an inclusive creative community where algorithm developers, users, and contributors to data pools all play vital roles in the creative process, democratizing and expanding the scope of artistic and architectural output (Melis et al., 2024).

Generative AI empowers artists and designers to integrate their work with their peers, facilitating artworks that exceed the cognitive capabilities of the human mind. This approach fosters a collective form of creativity that utilizes shared algorithms and databases, challenging entrenched beliefs about art and ownership. It advocates for a renewed perception of creativity as the result of both intentional and spontaneous collaborative efforts. This perspective underscores the communal nature of creative work, suggesting that our appreciation of art and design should evolve to include the contributions of generative AI as a vital and enriching aspect of the creative world.

The influence of generative AI extends seamlessly into the realm of performative arts, where it transforms the performer-audience dynamic and creates new immersive experiences. Just as in visual arts and architecture, AI's role in performative arts involves the categorization and analysis of vast datasets to generate creative outputs that challenge traditional notions of authorship and performance. This collaboration between human creativity and AI algorithms extends the scope of what is considered performance art, suggesting a reevaluation of the roles of the performer, audience, and medium in the creative process.

Generative AI's ability to analyze and synthesize data in real-time allows for dynamic and interactive performances that adapt to audience responses and environmental conditions, enhancing the immersive experience. This technological integration promotes a participatory approach to performance art, where the audience becomes an active contributor to the creative process, blurring the lines between creator and spectator. By incorporating AI into performative arts, artists can explore new dimensions of expression and engagement, fostering a more inclusive and dynamic artistic community.

The transformative impact of generative AI on art, architecture, and performative arts underscores the need for a reflective and adaptive approach to creativity. By embracing the collaborative potential of AI and expanding our understanding of creativity to include the contributions of artificial intelligence, we can unlock new possibilities for innovation and expression, fostering a future where human and machine ingenuity converge to advance artistic and societal progress.

Generative AI and Exaptation in the Creative Arts

Generative AI exemplifies the concept of exaptation, a term first introduced in evolutionary biology to describe how features initially developed for one purpose can be repurposed for entirely new functions. Stephen Jay Gould and Elisabeth Vrba brought this term into prominence in 1982, offering it as an explanation for evolutionary developments not easily explained by traditional adaptation narratives. Unlike adaptations, which are direct responses to environmental pressures, exaptations involve repurposing traits for functions they were not originally selected for (Gould & Vrba, 1982).

For instance, bird feathers initially evolved possibly for thermal regulation but later became instrumental in flight. This classic case of exaptation highlights how the original purpose of a feature can be overtaken by a new, advantageous function. This concept extends beyond biology to technology, architecture, and the arts, where existing tools, ideas, and practices find new life and applications unforeseen by their original creators.

In the technological sphere, the internet's evolution exemplifies exaptation. Originally designed for specific scientific and military communications, it has transformed into a global platform for various activities, including commerce, social interaction, and entertainment, far exceeding its original scope. Similarly, in the creative arts, repurposing materials or concepts to generate artworks that significantly depart from their origins underscores the inherent creativity in seeing potential beyond intended uses.

Exaptation thus highlights the dynamic and innovative nature of evolution and design, emphasizing the adaptability and potential for existing elements to be reimagined and repurposed. This understanding enriches our perspective on development and innovation, suggesting that progress often leverages what already exists in new and transformative ways.

In the creative and technological domains, generative AI's application of exaptation is particularly compelling. It involves the innovative use of existing data—images, sounds, or text—to produce novel artistic or design outputs. This mirrors the evolutionary process of adapting features for new functions, showcasing the capacity for what might be deemed obsolete or redundant to become the cornerstone of creative breakthroughs.

Creativity, at its heart, is about combining and reinterpreting existing elements to forge new, valuable creations. This process, intrinsic to both human and AI-driven creativity, depends on merging diverse ideas or concepts to produce original outcomes. The uniqueness of a creative work often lies in its distinct departure from original influences, achieved through blending and associative thinking. This challenges traditional views on authorship, rooted in the desire for individual recognition, and highlights creativity as a collective endeavor enriched by various contributions.

Generative AI's role in creativity transcends simple mimicry or reproduction. Instead, it embarks on a complex, sometimes autonomous journey of creating new combinations from existing data, akin to the human brain's idea generation through subconscious connections. This unpredictable, emergent quality of creativity allows for the development of truly innovative ideas and artifacts, underscoring generative AI's potential to produce genuinely novel contributions to art, design, and beyond.

The advancements in generative AI invite us to rethink our understanding of creativity, authorship, and the dichotomy between the natural and artificial. By recognizing AI's role in the creative process as an augmentation of human ingenuity rather than a replacement, we open the door to a more comprehensive and inclusive appreciation of creativity. This approach not only expands the realm of what is deemed creative but also fosters a synergistic collaboration between human and machine intelligence. As we delve deeper into the capabilities of generative AI, it becomes evident that the future of creativity hinges on our ability to integrate these technologies in ways that enhance and amplify our natural creative instincts.

Generative AI in performative arts can repurpose data and algorithms initially designed for other purposes to create innovative performance elements, such as real-time visual effects, adaptive soundscapes, and interactive narratives. This application mirrors the evolutionary principle of exaptation, where existing technologies find new, creative functions that enrich the artistic experience.

By integrating AI into performative arts, artists can explore new dimensions of expression and engagement, fostering a more inclusive and dynamic artistic community. This approach challenges traditional boundaries and encourages collaborative creativity, where AI and human performers work together to produce unique, engaging performances that reflect a blend of technological and artistic ingenuity.

Neomaieutics: The Generative AI Revolution in Performative Arts

The deployment and examination of Generative AI across diverse fields signify a seismic shift in performative arts, encompassing digital art, architecture, design, and beyond. These explorations challenge entrenched views on authorship and creativity, promoting the principle of exaptation as a cornerstone for groundbreaking innovation. Neomaieutics, derived from the Socratic method of eliciting knowledge through questioning, emphasizes the collaborative and iterative nature of learning and creation. Rooted in Socratic teachings, neomaieutics believes that knowledge is latent within individuals and can be brought to light through guided dialogue. This approach is embodied by the Verrocchio Workshop, part of the curriculum enhancement at the New York Institute of Technology, which fosters continuous pedagogical research bridging art, technology, and education. The workshop draws inspiration from Andrea del Verrocchio, the Renaissance master known for his collaborative studio that nurtured great artists like Leonardo da Vinci.

Within this neomaieutic framework, the installation "Two Acrobats" by Fadhil Fadhil and Monica Battistoni, showcased at FuoriSalone di Milano in 2023 with support from the IDC Foundation, exemplifies the innovative potential of Generative AI in performative arts. By blending the physical and digital, this work forges new dimensions in performance art, enriching the narrative of acrobatic performance with themes of resilience and theatrical enchantment. This initiative demonstrates AI's capacity to enhance the creative process and foster novel artistic expressions.

Similarly, the "Padiglione della Scienza" initiative, a collaboration featuring Emanuele Lisci and Dustin White, integrates AI-generated imagery to create a symbiotic bond between organic and engineered components within architectural design, showcasing Italy's dedication to technological progress and ecological stewardship.

Further exemplifying neomaieutic principles, the Italian Institute Culture New York's poetic creativity endeavor, spearheaded by Fabio Finotti, marries poetry with visual arts through AI-enabled image generation and 3D fabrication techniques. This project transforms abstract literary themes into tangible art pieces, expanding the boundaries of artistic exploration and highlighting the fusion of textual and visual creativity in performative arts.

Extending beyond specific tools like Midjourney or Stable Diffusion, the broader concept of collective creativity emphasizes the creative reuse of materials and functional co-optation, blending digital and analog realms through the collective exploration of ideas, images, and forms in performative arts. This approach, where intentionality becomes secondary to the collaborative process, is exemplified by the work of Heliopolis 21 Architects. Their practice-based research delves into architectural exaptation, enriching architectural taxonomies with eco-conscious strategies. Projects such as Cyberwall I and II, presented at the 2021 Venice Biennale Italian Pavilion, showcase sustainable and inclusive design by repurposing graphic compositions onto ceramic surfaces using Iris Ceramica Group's technology. These projects not only highlight innovative design possibilities but also spark discussions on the interplay between man-made and natural elements.

Eco-conscious Innovations in Performative Arts

The sustainable aspects of these projects are underscored by the use of Iris Ceramica Group's Active Surfaces®, featuring eco-active materials that combat pollution and promote health and environmental well-being. Installations like Spandrel II and Genoma at the Venice Biennale advocate for ecologically integrated architecture, combining aeroponic cultivation and biodiversity preservation, thereby fostering interdisciplinary collaboration and highlighting the need for creative and experimental architectural approaches to future challenges in performative arts.

In the 1970s, Ilya Prigogine won the Nobel Prize in Chemistry for elucidating a corollary of the second law of thermodynamics, stating that entropy determines the crisis of isolated systems. Prigogine discovered that every open system, which exchanges energy with the outside, reorganizes itself at a higher level of complexity. The climate crisis, due to the Earth's energy exchange with the Sun and the Earth's crust resulting in increasing disorder (entropy), is an open system that will reach equilibrium through a higher level of complexity (Prigogine, 1978).

This reflection dominated the idea behind the Genoma and Spandrel installations in the Italian Pavilion at the 2021 Biennale. Since entropy is irreversible, once the closed system enters into crisis, it is not possible to trace its causes back to closed models that, moreover, have caused the crisis. In the short term, it is not human physiology, which also undergoes this entropic process (which in this case we call evolution), that can become an open system consistent with the entropy of the Earth's energy system. But architecture necessarily must become an open system capable of reconfiguring itself at a higher level of complexity. This means considering the unpredictable, for now, consequences of the Earth's energy system exchange and feedback phenomena. I am not speaking of a voluntary choice on the part of the designers but of the irreversibility of the second law of thermodynamics, which no one can oppose.

Black Box and Createch installations further investigate the possibilities of architectural exaptation, exploring sustainable food production and the educational potential of innovative design practices. These case studies illuminate the transformative power of design and underscore the critical role of sustainability and multidisciplinary collaboration in shaping the future of performative arts.

The discourse on creativity often misconceives it as an individual endeavor, overlooking the intricate web of influences shaping each creative act. This view neglects the collective nature of creativity, which involves reconfiguring pre-existing ideas and elements. The dialogue surrounding authorship and AI integration in creative processes necessitates a reconsideration of creativity as a communal and iterative journey. By acknowledging the collaborative essence of creativity and the complementary role of AI, we can embrace a future where creativity flourishes through human and artificial intelligence synergy, fostering a more inclusive and innovative exploration of our collective creative potential.

In summary, the Verrocchio Workshop's neomaieutic approach to architectural education as an art performance highlights the profound impact of Generative AI on redefining creativity and authorship in performative arts. By integrating AI into creative practices, we can unlock new possibilities for innovation and expression, ensuring that technological advancements enhance the symbiotic relationship between human and artificial creativity.

Conclusion

In this chapter, we revisited our initial discussion on the conventional categories in art, highlighting the philosophical implications of reification and taxonomies, especially in times of global crises. Instead of reinforcing conventional categorizations, we challenged and expanded them, focusing on the transformative potential of Generative AI in performative art.

Generative AI emerged as a revolutionary force in the arts, disrupting established categories and fostering a new era of blended creativity and holistic perspectives. This technology enabled a collaborative relationship between human creativity and algorithmic computation, expanding the boundaries of what was considered art. By challenging traditional notions of authorship and creativity, Generative AI revived collective and performative art forms, prompting a re-evaluation of the roles of the artist, audience, and medium in the creative process.

Generative AI encompassed algorithms and systems capable of producing new, original content based on learned patterns and data. This creative process was driven by machine learning models, particularly generative adversarial networks (GANs) and variational autoencoders (VAEs), which iteratively refined their outputs through a feedback loop, minimizing the difference between the generated work and human-created work. This collaborative dynamic between human artists and AI algorithms democratized art creation, enabling individuals without formal art training to participate in artistic endeavors.

We concluded the chapter with final case studies showcasing the potential of this integrated approach, providing a glimpse into what this new paradigm of performative art could look like. These examples illustrated the dynamic interplay between human intention and algorithmic randomness, resembling a performance where the artistic process was as significant as the final product. Live generative AI installations, where algorithms created art in real-time in response to environmental inputs or audience interactions, further enhanced this performative aspect, making the creation process an integral part of the art itself.

Using these insights, we will outline, in the following section, a concise guide, serving as an initial discussion for a future foundational handbook for performing arts. This guide is intended to be shared in pedagogical and educational settings and to support scholarly research and professional practice in the fields of performing arts and multidisciplinary studies. By embracing the collaborative potential of Generative AI and expanding our understanding of creativity to include the contributions of artificial intelligence, we unlocked new possibilities for innovation and expression. This approach fostered a future where human and machine ingenuity converged to advance artistic and societal progress.

Guidelines for Designing Performative Spaces with Generative AI

1. ***Embrace Digital and Virtual Technologies:***
   (a) Utilize digital technology to create performance spaces that transcend traditional architectural constraints.
   (b) Incorporate virtual reality (VR) and augmented reality (AR) to enhance the performer-audience dynamic.
   (c) Design immersive experiences that offer new ways for audiences to engage with performances.
2. ***Foster Collaboration Between Human and AI Creativity:***
   (a) Implement generative AI algorithms to co-create content with human artists.
   (b) Utilize generative adversarial networks (GANs) and variational autoencoders (VAEs) to produce novel artistic outputs.
   (c) Encourage a symbiotic relationship where AI complements and enhances human creativity.
3. ***Redefine Authorship and Creativity:***
   (a) Challenge traditional notions of authorship by recognizing AI as a co-creator.
   (b) Promote a collective approach to creativity, involving both artists and AI in the creation process.
   (c) Emphasize the process of creation as much as the final product, valuing the dynamic interplay between human intention and algorithmic randomness.
4. ***Enable Real-Time Interaction and Adaptation:***
   (a) Design installations where generative AI can create art in real-time, responding to environmental inputs or audience interactions.
   (b) Develop adaptive systems that modify the performance based on audience feedback and real-time data.
   (c) Create performative spaces that allow for spontaneous and unpredictable artistic expressions.
5. ***Democratize Access to Art Creation:***

(a) Use AI to enable individuals without formal art training to participate in the creative process.
(b) Design tools and platforms that make generative AI accessible to a broader audience.
(c) Foster an inclusive environment where diverse community members can contribute to performative arts.

6. ***Integrate Educational and Pedagogical Elements:***

   (a) Develop a foundational handbook for performing arts that includes insights from generative AI practices.
   (b) Share guidelines and best practices in educational settings to support the development of new artists.
   (c) Use performative spaces as educational tools to demonstrate the potential of AI in art creation.

7. ***Focus on Interdisciplinary Collaboration:***

   (a) Encourage partnerships between artists, designers, technologists, and researchers.
   (b) Foster interdisciplinary projects that explore the intersection of art, technology, and society.
   (c) Use performative spaces as platforms for collaborative innovation and experimentation.

8. ***Promote Sustainability and Ecological Awareness:***

   (a) Incorporate sustainable design principles into performative spaces.
   (b) Use AI to optimize resource use and minimize environmental impact.
   (c) Design installations that highlight ecological themes and promote environmental stewardship.

9. ***Explore the Potential of Exaptation in Design:***

   (a) Apply the concept of exaptation by repurposing existing technologies and materials in innovative ways.
   (b) Foster creativity by reimagining the uses of tools and elements beyond their original functions.
   (c) Encourage a dynamic and adaptable approach to design, inspired by evolutionary principles.

10. ***Highlight Case Studies and Real-World Applications:***

    (a) Draw inspiration from installations that blend physical and digital elements to create new performance dimensions.
    (b) Explore projects that integrate AI-generated imagery with architectural design for symbiotic relationships.
    (c) Learn from initiatives that combines AI-enabled image generation and 3D fabrication with literary themes.

References

Akbarzadeh, M., Hablicsek, M., & Guo, Y. (2018, July). Developing algebraic constraints for reciprocal polyhedral diagrams of 3d graphic statics. In *Proceedings of IASS Annual Symposia (Vol. 2018, No. 16, pp. 1–8)*. International Association for Shell and Spatial Structures (IASS).

Berger, P. L., & Luckmann, T. (1966). *The social construction of reality: A treatise in the sociology of knowledge*. Anchor Books.

Bostrom, N. (2014). *Superintelligence: Paths, Dangers, Strategies*. Oxford University Press.

Ciftcioglu, O., Bittermann, & M. S., Yavuz, E. (2019, July). A Fuzzy Neural Tree for the Evaluation of Shape in an Architectural Design. In *International Conference on Intelligent and Fuzzy Systems* (pp. 238–246). Cham: Springer International Publishing.

Crawford, K. (2020). The trouble with bias. *Feminist AI: Vision and practice* (pp. 125–139). Carnegie Mellon University.

Elgammal, A., Liu, B., Elhoseiny, M., & Mazzone, M. (2017). *CAN: Creative Adversarial Networks, generating "art" by learning about styles and deviating from style norms*. arXiv preprint. arXiv: 1706.07068

Foucault, M. (1966). *The order of things: An archaeology of the human sciences*. Gallimard (Original French edition), Pantheon Books (English translation).

Foucault, M. (1977). *Discipline and punish: The birth of the prison*. Vintage Books.

Foucault, M. (1978). *The history of sexuality, Volume 1: An introduction*. Pantheon Books.

Gould, S. J. (1996). *The mismeasure of man*. W. W. Norton and Co.

Gould, S. J., & Vrba, E. S. (1982). Exaptation—A missing term in the science of form. *Paleobiology, 8*(1), 4–15.

Harari, Y. N. (2015). *Sapiens: A Brief History of Humankind*. Harper.

Huxley, T. H. ([1868] 2010). A liberal education; and where to find it: An inaugural address. In *Lay Sermons* (pp. 31–59). Cambridge University Press.

Kolarevic, B. (2019). Architecture of Change: Towards Transformable Architecture. Routledge

Liapis, A., Yannakakis, G. N., & Togelius, J. (2018). Computational game creativity. *Artificial and computational intelligence in games* (pp. 41–63). Springer.

Lukács, G. (1971). *History and class consciousness: Studies in Marxist dialectics*. MIT Press.

Manovich, L. (2001). *The language of new media*. MIT Press.

Marx, K. (1867). *Capital: A critique of political economy* (Vol. 1). Penguin Classics.

Melis, A., Pievani, T., & Lara-Hernandez, J. A. (2024). *Architectural exaptation: When function follows form*. Routledge.

Prigogine, I. (1978). Time, structure, and fluctuations. *Science, 201*(4358), 777–785.

Russell, S. (2019). *Human compatible: Artificial intelligence and the problem of control*. Viking.

Russell, S., & Norvig, P. (2016). *Artificial intelligence: A modern approach*. Pearson.

Veloso, P., & Krishnamurti, R. (2020). An academy of spatial agents: generating spatial configurations with deep reinforcement learning. *ECAADE 2020: anthropologic-architecture and fabrication in the cognitive age*, 191–200.

Wilson, E. O. (1984). *Biophilia*. Harvard University Press.

Yan, W. (2020). Computational Methods for Deep Learning: Theoretic, Practice and Applications. Springer.

MIX
Papier aus verantwortungsvollen Quellen
Paper from responsible sources
FSC® C105338

If you have any concerns about our products,
you can contact us on
ProductSafety@springernature.com

In case Publisher is established outside the EU,
the EU authorized representative is:
Springer Nature Customer Service Center GmbH
Europaplatz 3, 69115 Heidelberg, Germany

Printed by Libri Plureos GmbH
in Hamburg, Germany